New Directions in Philosophy and Cognitive Science

Series Editor
Michelle Maiese
Emmanuel College
Boston, MA, USA

This series brings together work that takes cognitive science in new directions. For many years, philosophical contributions to the field of cognitive science came primarily from theorists with commitments to physical reductionism, neurocentrism, and a representationalist model of the mind. However, over the last two decades, a rich literature that challenges these traditional views has emerged. According to so-called '4E' approaches, the mind is embodied, embedded, enactive, and extended. Cognition, emotion, and consciousness are not best understood as comprised of brainbound representational mechanisms, but rather as dynamic, embodied, action-oriented processes that sometimes extend beyond the human body. Such work often draws from phenomenology and dynamic systems theory to rethink the nature of cognition, characterizing it in terms of the embodied activity of an affectively attuned organism embedded in its social world. In recent years, theorists have begun to utilize 4E approaches to investigate questions in philosophy of psychiatry, moral psychology, ethics, and political philosophy. To foster this growing interest in rethinking traditional philosophical notions of cognition using phenomenology, dynamic systems theory, and 4E approaches, we dedicate this series to "New Directions in Philosophy and Cognitive Science."

If you are interested in the series or wish to submit a proposal, please contact Amy Invernizzi, amy.invernizzi@palgrave-usa.com.

Pii Telakivi

Extending the Extended Mind

From Cognition to Consciousness

palgrave
macmillan

Pii Telakivi
Practical Philosophy
University of Helsinki
Helsinki, Finland

New Directions in Philosophy and Cognitive Science
ISBN 978-3-031-35623-0   ISBN 978-3-031-35624-7 (eBook)
https://doi.org/10.1007/978-3-031-35624-7

Credit line: SEAN GLADWELL/Getty Images

This Palgrave Macmillan imprint is published by the registered company Springer Nature Switzerland AG.
The registered company address is: Gewerbestrasse 11, 6330 Cham, Switzerland

For my Grandfather,
who is no longer here
but who would always enthusiastically ask how many pages I had written

Acknowledgements

The first idea for this book was born when I started thinking about what would happen if I combined two of my favourite topics in philosophy of mind: phenomenal consciousness and the extended mind. I had just begun my PhD and was astonished to notice that nobody had done that research before in book length. Later, the dissertation evolved into this book. And times have changed also in the wider philosophical field: it is no longer an uncharted road but has gained growing attention and publications. I hope this book will further increase the interest and research about extended conscious mind.

This book would not exist without the encouragement and help of numerous colleagues, institutions, and funders. In particular, I am deeply grateful to Julian Kiverstein, Erik Myin, Alva Noë, Valtteri Arstila, and Gabriel Sandu. Julian gave me such an extensive list of comments that I could have written a second book based on them solely. He and Erik encouraged me to send the manuscript to a publisher, which was an important reassurance. Alva and the "Friday group" in his backyard in the Berkeley Hills offered the most inspiring philosophical environment to finalise the manuscript. I am also forever grateful to my alma mater, University of Helsinki, and the Departments of Theoretical and Practical Philosophy.

I am also greatly appreciative to audiences of various conferences all around Europe, and to all the colleagues who have given me insightful comments, new ideas, criticism, challenges, and support along the long process of writing this book. I especially want to thank Kristjan Loorits,

Samuli Reijula, Chiara Cappelletto, Laura Oulanne, Anna Ovaska, Saara Moisio, and Steven Gouveia for comments and invaluable advice. During the last few years, my academic home has been within the research group RADAR. I want to shout out my heartfelt thanks to Raul Hakli, Pekka Mäkelä, Tuomas Vesterinen, Tomi Kokkonen, Kamil Mamak, Olli Niinivaara, and Dane Gogoshin. I also want to send my profound gratitude to the editors at Palgrave Macmillan and to three anonymous referees, and to the Kone Foundation and the Fulbright Finland Foundation for financial support.

I am delightfully grateful to Louie the dog, who provided me with much-needed quadruped companionship while I was finishing the manuscript. I was also thrilled with a blue jay who came to check on me every day when I was writing in the Willard Park. And not to forget about my human friends in Berkeley and Helsinki—who patiently keep on reminding me that there is more to life than just writing philosophy books.

April 2023 Helsinki

Contents

Introduction: Out of Our Heads

In her newest novel *The Candy House*, Jennifer Egan creates a scenario set up in the near future, where people can upload contents of their consciousness online, share it with others, and search through other people's memories. The novel conjures up a thrilling (and somewhat dystopian) portrayal of humankind after the externalisation of consciousness has become possible.

The book you are reading now tells a very different story. Consciousness will be externalised in this book as well, but there is nothing dystopian, not even sci-fi about it. In fact, externalisation is a very mundane phenomenon, something we are all familiar with. The main principle is simple and doesn't require sophisticated modern technology. We have already extended many of our conscious and cognitive functions to diaries, smartphones, sketchbooks, and so on. Of course, more advanced technologies open up completely new ways to employ externalisation. In the course of this book, we will encounter sensory substitution devices, robotic fingers, and belts that create a feeling of the magnetic North, for example. After all, all comes down to the human capacity to skilfully integrate and incorporate external tools into themself.

Philosophers Andy Clark and David Chalmers asked a famous question in their *Analysis* paper in 1998: *Where does the mind stop and the rest of the world begin?* Twenty-five years later, their question is as pressing as ever. Our everyday tools and technologies are so deeply intertwined with us that it is difficult to disentangle the functions of our minds and the ones we perform with phones. Our smartphones "know" us better than our friends, spouses, or even ourselves, at least where the quantity of information is concerned. Many of the functions that we used to do "in our heads" only a decade or two ago are currently being extended to combine not only our brain functions but also our smartphones and other similar devices. Smartphones have become part of our *minds*.

This kind of "extended" view of *cognition* has become very popular in philosophy, cognitive science, and across many other disciplines over the last couple of decades. It has created a completely new paradigm and a way to conceive the human mind. However, in most of these accounts, consciousness is left out. This is where this book lands. The most important argument of this book is that if we accept that cognitive states can extend, we must accept that conscious experience can also extend. It makes explicit the argumentative path and shows in various ways, including through a number of empirical examples, that if we accept that cognitive states can extend, it is very difficult to keep consciousness in the head. Another important goal is to weave the motley, scattered threads related to extended consciousness together, into one book. The aspiration is to speak both to enactivists and to analytical philosophers of mind and hopefully bring them closer together.

In a broader sense, the central message of this book is that the human mind cannot be explained by focusing solely on what goes on in the brain. Extending the scope of the human mind and conscious experience has wide-ranging consequences for many fields inside and outside of philosophy, from cognitive sciences, psychology, and psychiatry to ethical and legal domains, technology research, artificial intelligence, and developmental robotics.

Finally, let me say a few words on how all this is structured in the book. The first chapter provides an overview of the theoretical background of externalism and 4E-theories in the present-day philosophy of mind and cognitive sciences. It sheds light on the central concepts, accounts, distinctions, and methodological questions that will be used and further developed in later chapters. It lays a path from Clark's & Chalmers's theory of *Extended Mind* (EM) to the position defended in this book, *Extended Conscious Mind* (ECM). In this chapter, this position is located at the top of *third-wave externalism* and as an attempt to bridge the hypotheses of extended mind and enactivism.

The second chapter presents three arguments for the position defended in this book. The general idea behind them is that by accepting three features of the 4E-framework, we need to accept ECM as well. The arguments build bridges from the hypothesis of extended cognition to ECM, from sensorimotor enactivism to ECM, and finally from the embodiment thesis to ECM. Most of the proponents of EM have not even considered the possibility of ECM, while some, most notably Andy Clark, have been arguing strictly against it. This chapter shows why his position is untenable and why accepting extended cognition leads to extended consciousness.

The third chapter analyses the lively ongoing debate at the intersection of philosophy of mind and philosophy of science about the causal–constitution distinction by going deeply into the concept of constitution. This analysis is necessary for answering two central critiques that have been raised against ECM, namely, the objections based on the *causal–constitution fallacy* and the closely related *cognitive bloat argument*. These critiques deserve their own chapter because it is crucial to show that the external elements are constitutive parts of the experiences under scrutiny and are not merely causally affecting them. The main argument of this chapter is that rather than interpreting it in mechanistic terms, the causal–constitution distinction should be interpreted in diachronic terms when dealing with mental phenomena. When so depicted, the extension relation in ECM counts as constitutive.

The fourth chapter sets the preconditions and limits for ECM. It distinguishes between four degrees of extension that build up a continuum: from more fleeting and temporary tool use to a more robust, integrated, and incorporated relationship with the device. The taxonomy reveals differences in the degrees of attachment and depicts what is extended: a certain distinct experiential function or the experiential subject as a larger system. The strongest of these degrees (what I call *functional incorporation*) is required for ECM: the external tool must become part of the transparent bodily point of view and under the sensorimotor control of the subject. Based on the notion of functional incorporation, a set of demarcation criteria ("the glue & trust criteria") for ECM will be developed. This chapter closes with a discussion of sensory substitution technology as a concrete example of a phenomenon that fulfils these criteria, that is, as a case of ECM.

The fifth chapter examines and answers four counter-arguments that have been set forth against ECM. They are (1) the argument from high bandwidth, (2) the argument from predictive processing, (3) the brain-in-a-vat thought experiment, and (4) the dream challenge. The first two counter-arguments aim to block the inference from EM to ECM, that is, the first argument for ECM presented in the second chapter of this book. The last two criticisms especially challenge the second argument for ECM (the inference from sensorimotor enactivism to ECM) but also the third argument (the inference from embodiment to ECM). However, this chapter shows that they are not successful in cutting off the inference relations of these arguments, and hence, none of them refutes ECM.

The concluding chapter connects the themes from all the preceding chapters, summarises key findings, and places them in a broader landscape that goes beyond philosophy of mind. It highlights the significance of this research by connecting the findings to ethical and societal questions that are particularly pressing in our everyday lives. This chapter also opens the door for future research questions and suggests where to go next.

CHAPTER 1

A Roadmap from the Extended Mind to the Extended Conscious Mind

1.1 Where Does the Mind Stop?

Technology has become increasingly prevalent in everyday life, reorganising how we act, think, and feel, and what we are. We use instruments, apps, and gadgets for various purposes, and oftentimes it is difficult to distinguish the tool from its user. Interfaces between the person and the tool, and the mind and the world, are obscured in many everyday situations, and this raises a great deal of questions. This book will present a view of the human mind that extends its borders to encompass the tools and technologies we use and will answer questions that are related, especially, to extended perceptual experience.

Andy Clark's and David Chalmers's famous question about the borders of the mind led them to argue that the frontier is not always tied to the skin and skull. The material basis of cognitive states can extend to the tools and instruments we use. In their view, the interface between the tool and the user has shifted—instead of two separate systems, we should consider them as a hybrid, coupled system. However, they limit the scope of the extension rigidly to cognition and exclude consciousness. Most philosophers who are sympathetic towards cognitive extension have agreed with them and claimed that even if cognition were extended, conscious experience would still remain in the head and be explainable in fully internalist terms.

P. Telakivi, *Extending the Extended Mind*, New Directions in Philosophy and Cognitive Science,
https://doi.org/10.1007/978-3-031-35624-7_1

In this book, I modify Clark and Chalmers's question and ask instead: *where does conscious experience stop and the rest of the world begin?* I argue that the skin and skull frontier is not always the answer, but sometimes the material realisers of conscious experience extend outside the bodily boundaries to include external elements. Extension happens with devices that are incorporated into bodily functions so that they have become part of the transparent perspective of the subject. Consider, for example, a blind person using a white cane to be able to move around. The cane has become part of her point of view: she does not primarily experience the cane in her hand but instead whatever is out there at the tip of the cane (what the surfaces feel like, etc.). I argue that the material realisers of these experiences are extended to also involve the cane and not only her neural functions. Hence, according to my position, neural functions are necessary but are not always sufficient for conscious experience.

Philosophy of mind and cognitive sciences have traditionally conceived mental processes as something that occurs in the head. Material realisers are strictly limited to neural functions and the central nervous system. However, in the recent couple of decades, an increasing number of philosophers have questioned the brain-boundedness of the mind and instead characterised mental functions as processes that include the body and the environment. Susan Hurley (2010, 104) asks, "what is so magical about the boundary around internal factors?" According to her, it is puzzling that philosophers of mind admit that they do not have the faintest clue about how functional properties could explain phenomenal qualities of experiences, but on the other hand, they are convinced that whatever the explanation is, it must be about neural factors. She calls this persistent internalist idea the *magical membrane* problem. One should admit, however, that there are no conceptual or logical requirements for internalism: "there is nothing about the notion of the subpersonal that restricts it to the neural in particular", as Dave Ward (2012, 733) reminds us.

Humans have always used various external tools and technologies to enhance, replace, and create new functions. Off-loading the cognitive burden in the environment is not a new phenomenon—even though it has taken striking forms with recent technological development. Already Plato wrote [sic] that *writing* works as an external memory for humans (in his *Phaedrus* dialogue). According to Clark (2003), we have always exploited environmental elements—incorporated and integrated material tools and cultural practices into ourselves. Biology has been intertwined with technology for a long time, and it is a completely "normal" situation for us. It is possible because of brain plasticity: our brains are sculpted by the tools we incorporate and by the activities we engage in.

The "E-Turn"[1] has become an auspicious line in philosophy of mind and cognitive sciences since the 1990s. It includes, for example, *enactivism* and the *embodied approach* (Varela et al. 1991), *dynamical systems theory* (Thelen and Smith 1994), *distributed cognition* (Hutchins 1995), *extended cognition* (Clark 1997; Clark and Chalmers 1998), and *sensorimotor enactivism* (Hurley 1998; O'Regan and Noë 2001). E-turn is a cluster of philosophical and scientific theories that reject the brain-boundedness of cognition. It is often referred to as the theory of the 4E's (*embodied*, *embedded*, *enacted*, *extended*). The roots of the E's lie, of course, deeper in history. Some of the most important earlier influences include James J. Gibson's (1979) ecological psychology and theory of *affordances*; Rodney Brooks's (1990) experiments in embodied, situated robotics; Jakob von Uexküll's (2010/1934) notion of *Umwelt* as describing subjectively produced surroundings; and Maurice Merleau-Ponty's (1945) phenomenology of the body. Obviously, the list could be extended, but that has to be left for historical surveys.

One single remarkably influential paper that has promoted the spreading of externalism is Andy Clark's and David Chalmers's "The Extended Mind" (1998), where they set the stage for a long-lasting discussion and debate.[2] According to their *extended mind* (EM) thesis, the material basis of *cognition* can be extended across the brain, body, and world. However, Clark (2009, 2012) and Chalmers (2008, 2017) deny that experiential states, such as perceptual experiences, could be extended. Many philosophers who are sympathetic towards cognitive extension tend to agree with Clark (and Chalmers) and leave conscious experience in the head. However, recent developments in the embodied, enacted cognition and sensorimotor approaches have created an appeal for a broader view. That

[1] A term coined by Dan Hutto.

[2] Before 1998, Clark had already introduced the idea of the mind extending outside the skin and skull in his 1997 book. However, the 1998 paper with Chalmers hit the jackpot. Even though this landmark paper has both names on it, the theory of extended mind is first and foremost Andy Clark's lifetime work. David Chalmers has stayed out of the extended mind debates after the original paper, apart from a few occasions: his foreword to Clark's book *Supersizing the Mind* (2008) and a short article, which was published in a collection of essays titled *Andy Clark and His Critics* (2017). The most recent comeback is made in his newest book about virtual and augmented reality (2022), but there he only recaps the original argument. It is noteworthy, however, that on all of these occasions he tries to deny ECM.

is precisely what I will do in this book. I will argue for a broader position according to which not only cognition but also consciousness can extend and call it the thesis of *extended conscious mind* (ECM).

One worry that arises is that if we accept that our smartphones (or some functions of them) are part of our minds, where does that take us? Is my laptop part of my mind, and what about the holiday pictures I have saved on it? What about Wikipedia or the whole of the Internet? Or consider pen and paper, a calendar, eyeglasses, a blind person's cane, a hearing aid, or a navigator. Or technologies and skills such as Braille writing, playing the violin, body hacking, augmented reality—or an old couple completing each other's sentences and looking back on their past from their shared point of view. Not all of these examples qualify as extended mind.[3] Clark and Chalmers's EM comes with instructions regarding when and where the extension can happen and which requirements need to be fulfilled. In this book, I will provide a similar treatment regarding ECM. I will give demarcation criteria of what kind of conditions need to be satisfied for something to count as an instance of ECM.

1.2 Concepts of Cognition and Consciousness

Let us now inspect two central notions that are used throughout this book: cognition and consciousness. There is no consensus amongst philosophers or cognitive scientists about the definition of either of the terms, where their limits are, or how they are materially realised. I do not think it is possible to give stringent definitions for these concepts in the first place, or that there is a *mark of the cognitive* or a *mark of the conscious* that would always lead to an infallible demarcation. (Adams and Aizawa's critique based on the mark of the cognitive will be discussed in the third chapter.) However, in what follows, I will outline some ways in which the concepts have been used in the sciences of the mind and explain how I will use them in this book.[4]

[3] This is related to one of the first counter-arguments that was presented against EM, namely, the *cognitive bloat argument*, which brings out the difficulty of demarcating the boundaries of extension. Especially the third and fourth chapters will provide responses to this critique.

[4] In regard to even more primary terms, I use 'mind' and 'mental' as synonyms, and they contain both cognitive and conscious processes. In other words, cognition and consciousness are subcategories of mind/mental. ('Mental' although is not the most advantageous term, as it might suggest a contrast with 'material'. This, however, is not my aim.)

Let us start with 'cognition'. It is a term used by many but defined by few. It is found in the titles of research fields (such as *cognitive* science and *cognitive* psychology), but the researchers in these fields might have varying ideas of its meaning. It is often said to contain at least dispositional beliefs, memory, attention, decision-making, reasoning, problem solving, using language, and other features connected with *thinking* in general.

Some take cognition as a *natural kind*, and others think it is better captured by focusing on *typical examples*. Cognition is *not* understood as a natural kind in this book. Instead, I follow the latter conduct: defining cognition by appealing to typical examples of cognition. However, this approach is not unproblematic either because what is considered "typical examples" varies significantly depending on the theoretical framework (see the discussion in Newen et al. 2018).

The central discrepancy (for the purposes of this book) lies between the *cognitivist* and *enactivist* understandings of 'cognition'.[5] According to the *cognitivist* reading, cognition is manipulation of representations, while the *enactivist* notion refers to the autonomous embodied agent's interactive relationship with her environment. The former has influenced how the concept is used in functionalism, which is a branch in analytic philosophy of mind, whereas the latter has to a great extent shaped how it is used in the 4E-framework. However, to make things more complicated, not all of the 4E literature uses the concept in the enactivist way (and not all analytic philosophy of mind is functionalism). In particular, it is crucial to note that the original extended mind account by Clark and Chalmers is *not* using an enactivist concept of cognition but rather a functionalist, cognitivism-influenced one. This situation embeds a tension within the research that encompasses both Clark and Chalmers's account and enactivism—from which the present study is not safe.

There is an ongoing debate whether accounts based on functionalism (such as Clark's & Chalmers's) can be fitted with enactivist accounts (for discussion, see, e.g. Kiverstein and Clark 2009). Both parties themselves have expressed sceptical views: proponents of extended functionalism (Wheeler 2010) and enactivism (Thompson and Stapleton 2009) have suspected the compatibility of the two approaches. On the other hand, there have also been more optimistic voices, such as Richard Menary's (2010b), who does not see an in-principle obstacle to combining the two

[5] Both *cognitivism* and *enactivism* are introduced in the next subsection, and are discussed in further details in later chapters.

views. As things stand, the jury is still out. This book is one attempt to bridge the gap and to bring the two standpoints closer.

Clark and Chalmers's argument was established within the functionalist conception of mind that understands cognition as something that is realised in the head. Without breaking out of the functionalist framework altogether, perhaps the only option was simply to move the physical barrier but not otherwise change the concept. Critical voices have said later (see, e.g. Hutto and Myin 2012) that this kind of framing repeats the internalist assumptions: in the default case, cognition is something that occurs in the head, and extensions are just exceptions that bring the normally internally realised functions into a wider realm. Even though I agree with the criticism that the way Clark and Chalmers (and many of their followers) use the concept of cognition does not reorient it as thoroughly as is needed, I still think the extension they made within the functionalist framework was highly valuable. The extension needed to start somewhere.

On the other hand, in enactivism, extension is not a deviation but is inherently embedded in the enactivist concept of cognition. In a way, it does not even make sense to talk about extension because in enactivism, the non-extended cases are the ones that count as abnormalities. In the enactivist account, what requires explanation is cognition that is *not* based on extension and interaction. For example, in the fifth chapter, I look at some enactivist ways to explain the occurrence of dreams (which is an often-used example of a phenomenon that at least prima facie seems internally realised, and hence exceptional and challenging for enactivists).

Some writers (such as Hutto and Myin) have tried to distance themselves from the functionalist reading of EM. My overall plan is the same, but unlike the typical enactivist writer, I take the functionalist reading seriously. I think this is the only feasible way when the starting point is Clark and Chalmers's framework, where the extension is understood in functionalist terms. I will, however, go beyond that framing and embrace the enactivist picture, which ultimately criticises the functionalist framework. Even though the two frameworks proceed to the matter from opposing directions—Clark and Chalmers's EM is trying to find arguments for why extension sometimes happens, and enactivism is trying to find explanations for why extension sometimes doesn't happen—they still share a common goal: a criticism of intracranialism. Despite the tension, I think the two frameworks can be used together for a common purpose. This book pursues extending the extended mind: leading the functionalist reading of EM to the enactivist sphere where consciousness comes along.

Another ambiguity between the two interpretations is that according to cognitivist reading, cognition is *not* experiential, while according to enactivist reading, it *is* experiential. As one of my central aims in this book is to show that Clark's EM (which deals with the non-experiential concept of cognition) leads to experiential extension (ECM), I have to acknowledge the way Clark uses these concepts. Hence, following Clark's notion, cognition is something that can be separated from consciousness, something *non-experiential*. The view that cognition and consciousness are separate has gained support from a commonly held view according to which *blindsight* is a neuropsychological condition where a person has cognitive states without being consciously aware of them. However, the consensus has been challenged; for example, Ian Phillips's (2021) recent work aims to show that blindsight is qualitatively degraded but still a conscious vision.[6]

Even though an interesting question, my account does not depend on whether there actually are non-conscious cognitive processes or not. However, there certainly are cognitive processes that are conscious (such as occurrent episodic memories). In practice, cognition and consciousness usually occur together, and it is often impossible to tell where one ends and the other begins. I will criticise the strict separation of cognition and consciousness in the second chapter (2.2). The critique concerns empirical occurrences of cognition and consciousness—not the bottom-line conceptual differentiation. The crux is that cognition and consciousness overlap, and they are often difficult to keep apart neatly in real-life situations. Nevertheless, there is a conceptual differentiation between them, and my aim is not to question that but instead to show that they are often intertwined with no gaps in between. (It is worth noting that the relation between the concepts of cognition and consciousness is not in a central role in Chap. 3, 4, or 5, nor in the second and third arguments in Chap. 2.)

Let us now turn to the concept of *conscious experience*.[7] If the concept of cognition was already multifaceted, consciousness is undoubtedly one of the most charged concepts in contemporary philosophy of mind. Scientists often try to avoid it altogether, and analytic philosophers of mind have created the most extraordinary ways of treating it—from bats to zombies. However, perhaps somewhat surprisingly, conscious experience

[6] I want to thank an anonymous referee for pointing out Phillips's work about the contentiousness of blindsight as an example of a cognitive, non-conscious phenomenon.

[7] I use synonymously the terms 'consciousness', 'conscious experience', 'phenomenal experience', 'experiential state', 'experiential process', and other versions of these terms.

is actually simpler to define than cognition since it covers a narrower area than cognition. In this book, conscious experience simply refers to *phenomenality*—that the mental state has a phenomenal side to it, that it feels like something to undergo it. It doesn't refer to the general capability of being a conscious creature or being awake and possessing awareness; in other words, it doesn't refer to *creature consciousness* (see Bayne 2007) or *access consciousness* (see Block 2002). Instead, the concept is used to refer to particular states or processes that are experienced in a certain something-it-is-like way.

The discrepancy between the cognitivist and enactivist accounts described earlier also affects how the concept of consciousness is understood. Within analytic philosophy of mind, starting from Thomas Nagel's (1974) famous paper, conscious experience has been understood as certain properties of "somethingness" it feels like to undergo an experience. This *what-it-is-like-ness* is often described in similar terms as the currently less fashionable notion, *qualia*. It conceives of experiential states as static, intrinsic properties: snap-shots of redness, painfulness, sweetness, coldness, and so on. Opponents in the same framework (eliminativists who deny the existence of qualitative states) also use the concept in the same way.

This understanding of consciousness faces trouble when combined with the concept of cognition as it is understood in the same framework. If cognition is understood in terms of computation and consciousness in terms of qualia, they are undoubtedly hard to fit in the same picture. Both sides (philosophers who want to stress the special nature of qualia and philosophers who want to eliminate qualia) end up separating consciousness from cognition. This has resulted famously in the *hard problem of consciousness* (Chalmers 1996) and the *explanatory gap* (Levine 1983). (See the discussion in the second chapter and Kiverstein 2016.)

On the other hand, when cognition is understood in terms of enactivism, consciousness is already on board. When cognition is not understood as computation and consciousness not as qualia, there is not a similar need for a strict separation. This is the reason why many enactivists state that the "hard problem" does not arise in the first place (see, e.g. Kiverstein 2016; Silberstein and Chemero 2015).

Furthermore, the explanandum of consciousness is not the same in *autopoietic* and *sensorimotor* enactivisms. While autopoietic enactivism describes consciousness as an autonomous agent's dynamically evolving relationship with its environment, sensorimotor enactivism focuses on *perceptual* experiences, which are based on our actions in the environment. (This subdivision within enactivism will be presented in more detail in the following section.) According to sensorimotor enactivism, perceptual experience is "something we do", as Alva Noë (2004, 1) famously wrote. Autopoietic notion is a way to explain the general capability of being conscious (as in *creature consciousness*), whereas sensorimotor enactivism deals with the capability of having perceptual experiences (as in *state consciousness*).

In summary, my use of 'conscious experience' refers to the phenomenal quality of mental states. It is the same feature that Nagel was referring to—even though I don't ascribe to all of the theoretical assumptions that lie behind the legacy of Nagel's conceptual work. The difference between my use and the analytic philosophy's use is that I reject the static atom-like nature that was encouraged by the notion of qualia. Furthermore, even though I follow Clark and Chalmers's conceptual framework for the sake of argument, overall my view is closer to enactivism's concept of cognition and consciousness. I criticise the cognitivist picture that understands cognition as computation, and this criticism gives the result that phenomenal experience doesn't have to be something extraneous. The target of ECM is *particular perceptual experiences*—processes such as reaching for an object, hearing a melody, or spatial navigation such as moving around in a room, namely, perceptual interactions that have a phenomenal quality in them.

To conclude, ECM is hard to accept for analytic (or functionalist) philosophers of mind who are sympathetic towards EM because it challenges the strict separation between cognition and consciousness. On the other hand, ECM might be trivial for enactivists who accept EM, since for them, cognition by definition already comprises experiential aspects. Hence, this book is valuable for both parties: for a functionalist philosopher, because it shows that consciousness cannot be cut out neatly from accounts of cognition, and for an enactivist, because it addresses an extension of phenomenality that has been taken for granted but never truly spelled out in explicit terms.

1.3 4E's: Embodied, Embedded, Enacted, Extended

In recent years, different sorts of externalist views in philosophy of mind and cognitive sciences have often been combined under the title '4E cognition'. Mind is *embodied*, that is, it is partly constituted by bodily processes; mind is *embedded* in its surroundings (Umwelt); mind is *enacted* through the agent's actions in the environment; and finally mind is *extended* to consist of external elements as its constitutive parts. The embeddedness thesis is a causal claim, whereas the extended thesis is a constitutive claim. The causal–constitutive distinction is the demarcation criterion between those two views. Embodiment and enacted theses are used in a less strict way in the literature: they are sometimes used as causal and sometimes as constitutive claims. However, throughout this book, I will treat them as constitutive theses. Thus, only one of the E's is causal (embedded), but in this book, if not otherwise specified, the rest of the E's always describe a constitutive relation. In the following, I will take a closer look at each of the 4E's.

1.3.1 Embodied

According to the *embodied* mind thesis, different aspects of the mind are intrinsically bodily phenomena or dependent on bodily functions. Depending on the writer, these aspects can include, for example, cognition (e.g. Gallagher 2005), perceptual experiences (e.g. Noë 2004), or affective states such as emotions and moods (e.g. Colombetti 2014). The body shapes, enables, and sets the limits for our minds. Its anatomy, structure, motor capabilities, and so on determine how we perceive the world and what we can do in it. The embodiment thesis can be weaker or stronger: it can be read as the mind being causally dependent ("enabled" or "affected") on bodily processes or constitutively dependent on ("made of") those processes. In this book, I follow a constitutive reading: the physical body and its functions are part of the constitutive basis for certain mental processes.

At least in principle, the embodiment approach can be accepted without the other E's. Consider the claim that *conscious experience* is embodied: technically, it is fulfilled when the body (and no other external processes) makes up part of the material basis of the experience in question. It would be *extracranial*, based on more than just brain functions, but the limit would lie there: the environmental elements outside the body would be left out.

Nevertheless, if one accepts the embodiment thesis, it is quite natural to continue to the other E's. The *embedded* and *extended* theses leak into the picture because it is difficult to set the limits of the body. A prosthesis, glasses, hearing aids, cardiac pacemakers, a blind's white cane, sensory substitution devices, and other non-biological devices can be incorporated into the body or into the bodily point of view—and as a result, the borders of the body become blurred. (There are different types of incorporation: for example, a cardiac pacemaker is a very different bodily extension compared to a blind cane. These differences will be analysed in Chap. 4.) Finally, the *enacted* thesis comes in because it is difficult, if not impossible, to differentiate the body from the actions it performs.

The embodiment thesis, on the other hand, is necessary for the other three E's. It is not possible to *embed* or *extend* in the environment without a body as part of the process. Nor is it possible to *enact* in an environment without a body—actions between the subject and her environment are necessarily embodied. Thus, the embodiment thesis is a gateway to the other E's.

The embodied approach will be discussed especially in the second chapter, where I argue that accepting the embodiment of experiences leads to ECM. It will also be discussed in the fourth chapter, where I differentiate between levels of embodiment and extension, from shorter-term uses of external tools to profounder cases of incorporation.

1.3.2 *Embedded*

According to the *embedded* thesis, the environment is necessary for the mind—the mind is "intimately embedded in its world" (Haugeland 1998) and profoundly "environmentally supported" (Sterelny 2010). Minds are never entities in a vacuum but creations enabled by interaction with their surroundings. The brain alone cannot support a mind, but the outer world is a necessary part of the package. Whereas in the embodied thesis mind was intrinsically a bodily phenomenon, in the embedded thesis, it is intrinsically an environmentally situated phenomenon.

The embedded thesis is a broader position than the rest of the E's. It is also "lighter", in the sense that many people who are otherwise sceptical about externalist views have accepted it (e.g. Rupert 2009). However, the embedded thesis is more than the trivial claim that the outside world affects our minds. Even an internalist could accept this much, for she could say that the mind depends on the sensory input from the world, but that its processing—the cognitive work—is done inside of the subject. The embedded thesis denies this and claims a stronger link. Instead, many of our cognitive processes are directly dependent on environmental interactions. In metaphysical terms, the embeddedness thesis is a causal rather than a constitutive claim (as opposed to, e.g. the extended thesis). The environmental interaction affects, influences, shapes, and enables the mind—but it is not part of its constitutive basis. Environmental elements are among the aetiological reasons how minds are made up. This is why the embedded thesis is easier to accept than the extended thesis.

The terms 'situated' and 'scaffolded' are sometimes used synonymously with 'embeddedness'. However, I follow Krueger and Colombetti (2015), who use *scaffolded* differently—not as a way to differentiate between causal and constitutive but simply as a way of stressing the role of environmental interaction without taking a stance on the causal–constitutive debate. They explain the benefits of a neutral term as follows. "As for the ontological question of when, if ever, the environment comes to be part of cognition—this is not something that a supporter of the scaffolded approach needs to provide an answer to" (Colombetti and Krueger 2015, 1159). However, the distinction between causal and constitutive is crucial for the position I am building in this book, and hence, the distinction between embedded and extended remains important. If used at all, by 'situated' and 'scaffolded', I always refer to the neutral stance that brackets out the causal–constitution question. I will discuss the outcome of the differentiation between the embedded and extended theses (i.e. the *causal–constitution debate*) in the third chapter.[8]

1.3.3 Enacted

According to the *enacted* thesis, our embodied actions and interactions in the environment form the constitutive basis of the mind. The term 'enactivism' was first presented in Varela, Thompson, and Rosch (1991). The account presented in their book is sometimes referred to as *autopoietic enactivism* by virtue of the central role of the *autonomous agent* that produces and maintains itself. All branches of enactivism, however, emphasise that the agent–environment relationship and interaction constitute a

[8] My position specifically addresses the constitutive components of the mind, not its causal background conditions. The distinction is important when discussing the metaphysics of mind. However, it is not so relevant in sciences, for example in psychiatry. Consider an Alzheimer patient, who relies on her own familiar home—the order of the furniture, other objects, and so on. Whether the home is a causal or constitutive element for her mental capacities is an irrelevant question for a physician, as long as it has a crucial effect. However, the constitutive dependency offers a stronger argument for protecting and appreciating the external aids.

mind. Let me give a short overview of this theory: what questions it aims to solve and why it is important for my argumentation in this book.

Enactivism emerged as a critical response to classical cognitive science, also known as *cognitivism*. Cognitivism is an intellectualist, internalist, computationalist, and representationalist view about cognition and mind (see, e.g. Fodor 1975). According to cognitivism, where there is cognition, there is computation. In the manner of digital computers, human cognition is also inherently the manipulation of mental representations: "the mind is thought to operate by manipulating symbols that represent features of the world or represent the world as being a certain way" (Varela et al. 1991, 8). The legacy of cognitivism in cognitive sciences and philosophy of mind has set the focus on brain functions and ignored the role of the body and environmental interactions (see, e.g. Newen et al. 2018).

Enactivism challenged all this by stressing the following features: 1) Living beings are autonomous systems that actively generate and maintain their identities. They don't just passively process pre-existing information from the world, but they *bring forth* or *enact* their own cognitive spheres in ongoing, continuous reciprocal interaction with their environment. 2) The nervous system does not process information in a computationalist way. It should not be understood as *hardware* but rather as an autonomous system on its own that actively generates its own meaningful patterns in interaction with a sensorimotor network. 3) Cognition is an embodied action. Cognitive processes are based on repeated sensorimotor patterns of perception and action.[9] 4) A cognitive agent's environment is not a predetermined or fixed area that is represented in its brain but rather a relational domain created by the mode of coupling with it. 5) Experience is not considered an epiphenomenal by-product, and the questions of cognitive science and phenomenology should not be kept separate from each other. (These features are presented in Colombetti and Thompson 2008, 55–56.)

Features 3, 4, and 5 are especially important for my argumentation in this book. As feature 3 (embodiment) was already discussed in the previous section (and will be discussed at length in later chapters) and feature 5 will be discussed in the next chapter, let us now focus on one central idea

[9] The third feature contains a stronger and a weaker claim. First, it states an identity ("cognition *is* embodied action"), and second, a causal or constitutive relation ("cognition is *based on* sensorimotor patterns"). This is a good example that both views can be found in the literature and that they often get confused.

of enactivism that is immersed in at least features 1, 2, and 4—namely, the concept of *sense-making*. It is a concept that I will use throughout this book.

Enactivism maintains that living systems "make sense" of their environments owing to their autonomous and adaptive nature. They care and worry about their world from their point of view—the environment is not indifferent to them, but it gets meaning by virtue of the sense-making relationship that is created in the interaction between the agent and their environment. The paradigmatically used example of sense-making is *E. coli* bacterium that swims towards higher sucrose concentrations and swims away from noxious substances. Sucrose is significant as food for the bacterium, and hence it has meaning and value from the point of view of the bacterium. Even simple organisms such as bacteria make sense of the world based on their own needs (Thompson 2007).

Sense-making has similarities with the concept of 'valence' in psychology: we value our surroundings and its objects based on their attractiveness or averseness. Without the meaning-creating point of view to the world and purposeful goal-drivenness generated by the sense-making attitude, there cannot be true agency or mind (but only reacting to stimuli, as programmed machines do). This interactive relationship with the environment is not only enabling the mind, but it *is* mind: the enacted thesis is always a constitutive thesis.

Apart from autopoietic enactivism as a general approach to the mind, two more focused branches of enactivism are often classified. *Sensorimotor enactivism* is specialised in perceptual experience (see especially O'Regan and Noë 2001; Noë 2004; Hurley 1998). According to it, perceptual experience is based on the agent's sensorimotor actions and skills (*sensorimotor contingencies*) in her environment. It is realised in the course of sensorimotor activity itself—perception and action are mutually dependent processes. Finally, *radical enactivism* (or *radical embodied cognitive science*) takes the anti-representationalist side further than the other versions of enactivisms. Thus, it is not only criticising the cognitivist tradition but also showing that autopoietic and sensorimotor approaches contain leftovers of representationalism.[10] According to the radical view, basic forms of mind are non-representational and content-free—mind is constitutively "world-involving".

[10] Chemero (2009); Silberstein and Chemero (2011, 2015); Hutto and Myin (2012, 2017).

In this book, when I use the term 'enactivism' in a general sense, it refers to all of these branches. However, autopoietic and radical versions are not directly related to perceptual experience, and hence, they are mostly left in the background. They are not in contradiction with ECM, but their focus is elsewhere. Sensorimotor enactivism, on the other hand, addresses exactly perceptual experience, which is why most of the enactivism discussed in this book considers the sensorimotor branch (it will be discussed at length, especially in the second chapter). However, I use the notion of *sense-making* (which is more commonly used in autopoietic than in sensorimotor enactivism) to define conscious experience.

1.3.4 Extended

The last of the E's is the *extended* thesis, according to which non-biological environmental objects—tools, devices, instruments, equipment, props, and so on—can become part of the constitutive material basis for certain mental states. It is the strongest and, in a sense, the most radical of the E's. It builds on the other three E's and seems to require all of them to hold. However, some philosophers think a shortcut is possible. Extended functionalists such as Michael Wheeler and Andy Clark embrace "the strongest E" alongside the embedded thesis, and to a certain degree the embodied thesis, but they think it is possible to leave out the enacted thesis (or rather enactivism as a theoretical framework).

However, in my account, the extended thesis actually requires all the other three E's to hold. It is the environmentally embedded, embodied, (en)acting mind that can be extended with the means of external tools. The interconnectedness and interdependency of the E's becomes clear in the next chapter.

What makes the extended thesis such a striking position is that it is a *constitutive* claim. The tools don't just *affect* mental processes (whether cognitive or experiential) but are part of what mental processes are *made of*. The most important work supporting this view is Clark and Chalmers's (1998) theory about extended cognition (EM), and of course, the central goal of this book is to show that ECM belongs to this group as well. Accordingly, the extended thesis is the common thread running through this work. All the chapters address the question of whether experience is *extended*.

Apart from these above described 4E's, some other notions are sometimes added to the list (even though they don't start with the letter 'E'). I think we should add at least two: *distributed cognition* and *participatory sense-making*. They both touch on an issue that is left out from the other E's, namely, social and intersubjective relations. The view about *socially distributed* cognition is based particularly on Edwin Hutchins's (1995) work: cognition is not confined to an individual but spread to other members of the social group as well. Intersubjectivity has been a neglected topic in the enactivist literature until recently.[11] *Participatory sense-making* is correcting this by shifting the focus to intersubjectively shared features of the mind: it drives the enactivist notion of *sense-making* towards the social domain. According to it, intersubjective processes (e.g. a conversation) also self-organise: subjects coordinate their movements, actions, gestures, and so on, and this embodied intersubjectivity allows them to participate in each other's sense-making.[12] I wholeheartedly agree with the role of intersubjectivity in cognitive and experiential processes. However, I don't discuss the social aspect in this book apart from a few occasional examples.

All the E's (and other letters) presented here are more or less related to ECM. However, these E-views form a general theoretical framework that operates with a much larger canvas than ECM. Instead, ECM concerns only a narrower area: the constitutive material vehicles of certain perceptual experiences. I argue that three of the E's lead to ECM: *extended*, *enacted*, and *embodied*. In the second chapter, I will present three arguments based on these hypotheses.

1.4 From the Extended Mind to the Extended Conscious Mind

Let us now turn to the more specific theories of extended mind and extended conscious mind. They both belong to the last of the above introduced E's (*extended*). In the following subsections, I will first introduce the original formulation of EM as it was presented by Clark and Chalmers (1998). Then, I will present three waves of research that illustrate how the

[11] As Shaun Gallagher (2008, 176) condemns: "[O]ne can read through Noë's book without bumping into anyone else (with the exception of referenced scholars and researchers, of course). The idea that there are other people in the world is almost completely absent from the analysis of perception".

[12] Participatory sense-making was developed by Hanne De Jaegher and Ezequiel Di Paolo. (See, e.g. De Jaegher and Di Paolo 2007; De Jaegher et al. 2010.)

theory of EM has developed during the last 25 years and that ECM is riding on the crest of the third wave. In the latter parts, I will discuss how ECM has been defined in the literature and formulate my own definition of it. Finally, I will attend to *vehicular* externalism, that is, explain what is meant by the claim that EM and ECM concern the material vehicles of cognition or consciousness, respectively.

1.4.1 The Extended Mind

The starting point for my position on ECM is Clark and Chalmers's (1998) account of extended mind.[13] Although some other theories, especially sensorimotor enactivism, are actually in many ways closer in their spirit to ECM than Clark and Chalmers's EM is, the latter is still the central inspiration behind ECM. Therefore, let us take a closer look at Clark and Chalmers's argumentation in their ground-breaking work.

Clark and Chalmers offer a delightful number of concrete examples of cognitive extension, of which the most illustrative ones are rotation in Tetris and the amnesiac Otto and his notebook. In the former, Clark and Chalmers (p. 7–8) ask us to consider three cases of problem solving. In the first scenario, a person is watching a game of Tetris on a computer screen and has to "mentally rotate" the shapes—only by looking at them try to decide which way the shapes would fit the slots. In the second scenario, a person is watching the same geometrical shapes falling, but now she can physically rotate them with a rotation button (as in real Tetris). In the third scenario, a person in the cyberpunk future is playing the same game, but this time she has a neural implant that takes care of the rotation.

The gist of this example is that all three of the scenarios are on a par regarding their functions. According to Clark and Chalmers, the third case (neural implant) is clearly similar to the first case (mental rotation). The second case (rotation button), on the other hand, fulfils the same function as the neural implant, although it is distributed across the player and computer instead of being internally realised within the player. Clark and Chalmers ask: if the rotation counts as cognitive in the case of the neural

[13] I use the term *extended mind* (EM for short) throughout this book to refer to Clark and Chalmers's theory of extended *cognition*. The term is undeniably slightly misleading, and many writers have recently started to refer to it as the *hypothesis of extended cognition* (HEC for short). However, I think EM works better in the context of this book, where it is constantly compared with ECM. The terms are used consistently—the reader just has to remember that EM does not refer to *mind* in general, but only to *cognition*.

implant, by what right do we count the case of the rotation button as fundamentally different? Pointing to the boundaries of the body would just beg the question, since that boundary was at issue in the first place.

Clark and Chalmers (p. 8) discuss an experiment conducted with Tetris players (Kirsch and Maglio 1994). Physical rotation proved to be significantly faster than mental rotation: getting a shape through 90 degrees with a rotation button took 100 milliseconds plus 200 to select the right button, whereas the same action with mental rotation took 1000 milliseconds. Kirsch and Maglio showed that physical rotation is used not only for positioning the shape to fit the slot but also as a means to determine if the shape and slot are compatible. They call this an *epistemic action*. Clark and Chalmers (p. 8) take the idea further and propose that epistemic actions deserve *epistemic credit*. Hence, it is fair to say that the cognitive action in the second case is spread to also include the physical rotation.

The second example is the most famous one, and it wouldn't be unfair to call it the flagship of EM. Otto suffers from a curious version of Alzheimer's[14] that has made him very forgetful. He writes down everything in the notebook he always carries with him, and that's how he is able to cope in society like any of us. One day, he wants to visit the Museum of Modern Art in New York, and as usual, he consults his notebook. He finds the address, walks to the destination, and enjoys the art experience. Inga, on the other hand, has normal memory, and she also appreciates modern art. She decides to go to the same museum as Otto. She thinks for a moment to remember the address, recalls that it is on 53rd Street, walks there, and enjoys herself as much as Otto. Clark and Chalmers (1998, 13) state that the two cases are on a par, considering the status of their beliefs.

> [I]n relevant respects the cases are entirely analogous: the notebook plays for Otto the same role that memory plays for Inga. The information in the notebook functions just like the information constituting an ordinary non-occurrent belief; it just happens that this information lies beyond the skin.

[14] It has nothing to do with real Alzheimer's syndrome. The Otto example is a thought experiment, where it only matters that his biological memory is non-existent, whatever the diagnostic reason is. In general, I think we should be suspicious of the use of thought experiments in arguments in philosophy of mind. However, this particular thought experiment is rather innocent and quite successful in its purpose. On the one hand, it is based on a common everyday tool (a notebook) rather than some science-fiction scenarios or imaginary future technologies, and on the other hand, it works perfectly as an intuition pump that directs even the unaccustomed readers away from internalism towards externalism.

The aim of this example is to show that what makes something count as a belief is the role it plays—rather than whether it is located inside or outside the skin. Otto and his notebook together create a "*coupled system* that can be seen as a cognitive system in its own right" (C&C, 8). Hence, the notebook not only influences Otto's cognitive process of remembering the address but also becomes part of that cognitive process itself. Clark and Chalmers list requirements that need to be fulfilled for something to count as an instance of EM: the external device has to be a *constant* for its user, and it has to be easily and directly available, reliable, and automatically endorsed. (These "glue and trust" criteria will be analysed in the fourth chapter.) In particular, the example of Otto and his notebook has evoked an abundance of literature in philosophy of mind and related fields—critical comments, answers to critiques, applications to other areas, and so on. Many of those critiques will be discussed (and answered) in this book.

1.4.2 The Waves to the Extended Conscious Mind

Externalist accounts have developed in various directions since the original argument was presented 25 years ago. Broadly speaking, research on the extended mind hypothesis has been classified into three different theoretical "waves" (see, e.g. Menary 2010b; Gallagher 2018). *The first wave* refers to the original formulation of EM by Clark and Chalmers that is based on functionalism—the *functional role* the external element plays in the system. When the tool plays the same functional role as the biological process, it merits the same cognitive status. This is called the *parity principle*, which Clark and Chalmers describe as follows:

> If, as we confront some task, a part of the world functions as a process which, *were it done in the head*, we would have no hesitation in recognizing as part of the cognitive process, then that part of the world *is* (so we claim) part of the cognitive process. (Clark and Chalmers 1998, 8)

The parity principle has been criticised by both the supporters and opponents of EM.[15] One shortcoming is that it might encourage one to

[15] The parity principle will be under scrutiny in Sects. 2.2, 4.2.2, and 4.3.1. Note also that the term 'parity principle' was not used in C&C's original article. It has been introduced later (e.g. in Clark's contributions in the collection of essays edited by Menary 2010a), and I think the demand for this term evolved exactly from the need to differentiate the first wave from the second-wave EM.

think that mental functions can only be found by comparing them with biological, neural functions. Otto's notebook is part of his cognitive system *if* it is on a par with Inga's organic cognitive system. However, this was not what Clark and Chalmers had in mind. As Michael Wheeler writes (2015b, 158, footnote 3), Inga's role as a point of comparison could even have been left out without the core of the argument suffering. In hindsight, without comparing Otto to Inga, the philosophical literature would have been saved from the debate of whether the two ways of having a memory are on a functional par or not. In a sense, it was an unfortunate path because the truth of EM is not dependent on whether Otto's and Inga's beliefs are functionally identical (it is clear that they are not) or even similar enough. On the other hand, the critique led many supporters of EM to move their focus from the original parity-based arguments to other kinds of arguments.

The *second-wave* arguments dropped the parity principle and replaced it with theses such as *complementarity* and *integration*. John Sutton (2010) initiated the *complementarity principle*, which allows that there are dissimilarities between external and internal functions. Biological and external functions can sometimes be on a par, but often the externalised function is in fact altogether different from a biological one—and can still count as cognitive. Hence, parity is no longer a defining feature of extension. Sutton also emphasises that extension varies based on differences in people, situations, and environments. Subjects might have different habits of using tools—differing from person to person, situation to situation, and environment to environment. Menary's (2007; 2010c) account of *cognitive integration* is another example of leaving the parity requirement behind, instead focusing on the everyday embodied ways of integrating external means to the cognitive functions we perform. Menary sums up the second wave: "the bodily manipulations of external vehicles are different from, but complementary to, internal processes" (Menary 2010c, 240).

Finally, the *third wave* is an attempt to coherently bring together all the different threads, "a motley connection of factors", that have stayed somewhat disconnected in the earlier waves of arguments—from neurons to cultural practices and from writing skills to smartphones. Whereas the first- and second-wave arguments are rather established, the third wave is still in formation (Gallagher 2018). Sutton (2010, 213) describes the third wave as rejecting the idea that internal and external, natural and artificial would have fixed properties, but instead "this third wave would analyse these boundaries as hard-won and fragile developmental and cultural achievements, always open to renegotiation". Instead of fixed properties, the elements are constantly transformed, negotiated, and reformatted.

Shaun Gallagher (2018) mentions two attempts that have been initiated in the 4E literature to better take into account the dynamic, holistic conception of mind/cognition. On the one hand, *predictive processing* offers a unifying explanatory model, and especially Andy Clark's (e.g. 2013, 2015, 2016) recent work has brought the theories of predictive processing and extended mind together. On the other hand, *enactivism* is a theoretical framework that has always understood cognition in a dynamical way. One defining feature of the third wave has been to try to bring enactivism and extended mind together. The two theories have much in common, but they also contain sharp differences, such as diverging attitudes towards functionalism and representations. This book fits precisely this latter slot—third-wave externalism that understands mind in a dynamic enactivist way combined with lessons from the theory of extended mind.

Indeed, we can say that ECM is on the top of the third wave (which has also been described as a tsunami).[16] As stated, it is also an heir of EM, which has its roots in the first wave. How is ECM different from EM then? Of course, many cognitive states are conscious, and most conscious states are cognitive. Hence, the difference between cognitive vs. conscious cannot be used as a straightforward separating factor (even though it seems that most of the critiques assume consciousness can be neatly cut out from cognition). One important difference is related to the question of whether the extended state is occurrent or dispositional. The Otto example is about a dispositional belief, and dispositional beliefs are not experiential; therefore, the Otto example is not an instance of ECM. (It is plausible that Otto goes through some sorts of experiential states while checking the address, but that is not what the example is about.) ECM applies to occurrent states only, because experiences are always occurrent. However, cognitive states can be both occurrent and dispositional.

Consider the two examples I presented above by Clark and Chalmers (1998). The first one, using the rotation button in Tetris, concerns an occurrent process, while the second one, Otto and his notebook, concerns

[16] Gallagher (2018, 434) writes: "The first two waves of extended mind theory, as I count them, are part of the tidal change of embodied cognition that, on some accounts, has led to a gradual erosion on the shoreline of the classic cognitivists. The third wave promises to be more tsunami like and threatens to wash away that shoreline altogether".

a dispositional state. Thus, the Tetris example comes closer to ECM than the Otto example. Even though originally intended to give support exclusively to EM, it ends up dealing with aspects that are not only cognitive but also experiential, namely, sensorimotor actions.[17] Playing Tetris also undoubtedly requires cognitive actions—it is a good example of an occurrent mental process that consists of both cognitive and experiential aspects. I will discuss the difference between dispositional and occurrent EM in more detail in the second chapter, arguing that accepting occurrent EM leads to ECM.

In the remainder of this section, I will frame a definition of ECM (and EM). First, let us examine how others have defined ECM. Andy Clark[18] (2009) describes ECM as follows: "the local material vehicles of some of our conscious experiences might include more than the whirrings and grindings of the brain/CNS" (p. 967) and "the minimal physical substrate for some forms of conscious experience include[s] the goings on in the (rest of the) body and the world" (p. 963). Michael Kirchhoff and Julian Kiverstein (2019, 25) state that "The thesis of extended consciousness claims that the biological machinery that realises conscious experience can sometimes include a mixture of neural, bodily, and environmental elements".

Dave Ward (2012, 732) understands "ECM as the claim that *the subpersonal basis*[19] *of some conscious experiences can include events and properties outside the organismic boundaries of the experiencing subject*". Ken Pepper (2014, 99) is more specific and stresses the sensorimotor dynamics of perceptual experiences: "Some authors have urged that if it is indeed the case that perceptual experience depends on a bodily engagement with the world, then the implicated patterns of environmental interaction—sensorimotor dynamics—are physical constituents of perceptual experience". Daniel Hutto and Erik Myin (2012, 158) take a stance from supervenience and write that "phenomenality has a wide supervenience basis—that it constitutively involves parts of the environment. [...]

[17] Notice that Clark and Chalmers (1998) do not treat the rotating actions as conscious. As said earlier, they treat them as *epistemic* actions. However, I think their treatment misses the experiential side that is present due to the sensorimotor nature of the task.

[18] Note that Andy Clark himself is not supporting ECM.

[19] Ward (2012, 733) explains his use of 'subpersonal': "I focus on the claim about the subpersonal rather than the material basis only to avoid commitment to the idea that the best way of characterising that subpersonal basis will be in physical terms (as opposed to, say, the language of ecological optics)".

Phenomenal experience strongly supervenes on, or is constituted by, temporally extended, interactive worldly engagement".[20]

What is common in these formulations is that they all concern the constitution of the physical underpinnings of phenomenal experience. These definitions can be summarised as follows: *The constitutive material vehicles (i.e., the material basis) of some conscious experiences can include external parts.* However, my thesis in this book is more specific and limited: it only includes certain kinds of *perceptual* experiences that are extended with certain kind of tool use. This is one major difference between this book and Michael Kirchhoff's and Julian Kiverstein's (2019) recent monograph, which argues for extended consciousness. In their account, material realisers can also be composed of cultural practices and other people. Another considerable difference is that the crux of their book is to use the theory of predictive processing developed in cognitive neuroscience to support ECM, whereas in this book, the predictive processing framework is presented in more moderate terms: the aim is to show that it is not in contradiction with the idea of ECM. Kirchhoff's and Kiverstein's book has thus far been the only monograph with a direct regard to extended conscious mind. Despite the differences, our books share the common intent: to show that the realising underpinnings of consciousness are wider than has been traditionally thought.

Furthermore, my position is not modal—it is not a thought exercise what *could* or *can* be the case—but rather it is a claim about the *actual* state of affairs. Finally, my position concerns larger experiential *systems* rather than single experiential states or processes. For example, it addresses the use of a blind person's cane as a stable, consistent way of perceiving the

[20] For example, in the following texts, the writers have explicitly argued for extended conscious experience: Kiverstein and Farina (2012), Kiverstein (2016), Kirchhoff and Kiverstein (2019, 2020), Noë (e.g. 2004), Hurley (e.g. 2010), Rowlands (2010), Ward (2012), Pepper (2014), Loughlin (2012), Vold (2015), Bruno (2013), Silberstein and Chemero (2011, 2015). Many writers are also very close to ECM but have not fully accepted it or haven't taken a stance about it. They include Hutto and Myin (2012, 2017), Colombetti (e.g. 2014), and Wilson (2004). Actually, probably the majority of the 4E-oriented philosophers belong to this group, which makes the central argument of this book (that ECM follows from most of the E's) noteworthy. Yet another option is put forward by Michael Wheeler (2015a, b), who remains deliberately agnostic about ECM. He criticises certain arguments for ECM presented in the literature, but in the end his aim is to show that those arguments do not yet resolve the case for either externalism or internalism. Wheeler's views will be discussed in various places in this book.

world, rather than focusing on one successful use of a blind person's cane by someone who hasn't incorporated it into her bodily point of view.

This distinguishes ECM from a common reading of EM, according to which the unit that is extended is not the subject as a whole but rather just some of the subject's cognitive states or processes. I think this reading is unsatisfactory, but it is commonly used in the literature. However, Clark and Chalmers themselves would probably not commit to it, since the Otto example clearly points to an extension where a larger system (Otto's memory system as a whole) is concerned. They have added a short passage at the end of their paper, where they contemplate whether an extended mind implies an extended self, and their answer is "It seems so". "Otto *himself* is best regarded as an extended system, a coupling of biological organism and external resources" (Clark and Chalmers 1998, 18). In this book, I follow this broader reading of EM, where instead of separate cognitive states, a larger systemic whole is the unit that extends. The differentiation between single short-term extensions and wider-ranging systemic extensions (or *incorporations*, as I will specify them) will be illuminated in the fourth chapter. Based on all these reflections, I will define ECM and EM as follows:

> **Extended Conscious Mind (ECM)**: The constitutive material components of certain perceptual experiences include external elements on the condition that those components are functionally incorporated into the subject.

> **Extended Mind (EM)**: The constitutive material components of certain cognitive states/processes include external elements on the condition that they are integrated into the subject.

These are the definitions that are used in this book for EM and ECM. It is worth noting already at this point that they also contain the demarcation criteria for both EM and ECM—*integration* in EM and *functional incorporation* in ECM. Incorporation is a deeper relation than integration: it requires a change in the phenomenological status of the tool. When a tool is functionally incorporated, it has become part of the transparent point of view of the subject. The differences between them will be discussed at length in the fourth chapter.

Furthermore, in principle, ECM could be taken as an ontological or explanatory account—at least from a "philosopher's point of view". Many philosophers would probably consider an ontological account stronger than an explanatory account, and accordingly only an ontological account would count as "real externalism". An explanatory account, for them, would be something less because it would not touch the metaphysical level. However, I do not think it is self-evident that the ontology of mental states can be separated from explanations of those mental states. Empirical scientists only deal with explanatory accounts to start with, and wouldn't consider them any weaker, since empirical scientists don't postulate a separate "ontological" level. Of course, some philosophers follow this latter way as well (such as Susan Hurley).

When our aim is to do cognitive science rather than metaphysics, I do not think it makes much difference whether ECM is considered an explanatory or ontological claim. In any case, my account of ECM is "the strong claim". However, I acknowledge that some philosophers would only consider ontological claims as "strong" claims—and hence for them I say that ECM is an ontological account, whereas for (cognitive) scientists I say it is an explanatory account (and for them it goes without saying that what philosophers call "ontological" is not anything extra but is already there).

1.4.3 Vehicular Externalism

As became clear from the above definitions, both EM and ECM concern the ***material vehicles*** of mental states (the former of cognition and the latter of consciousness). I use the term 'vehicle' synonymously with 'material basis', 'material underpinnings', 'material realiser', 'sub-personal basis', 'substrate', 'material component', 'constitutive component', 'material part', and other similar combinations modified from these. They are the underlying entities that realise mental states. In internalist accounts of mind, neural functions are considered to be the only constitutive vehicle, whereas EM and ECM add the body and environmental interactions. Needless to say, brain functions are still *necessary* vehicles—they just are not always *sufficient* according to externalist views.

The fact that EM and ECM concern the material basis separates them from some other well-known theories in philosophy of mind and language. EM and ECM differ from Hilary Putnam's (1975) *semantic externalism* in that the latter is about mental *content*. It is often stressed (e.g. Clark and Chalmers 1998; Menary 2010b) that EM is a form of *active externalism*, whereas semantic externalism is *passive*—about the distal (albeit external) causes that have determined the contents of certain mental states. While this distinction is legitimate, I think it dismisses the central disjointing factor. Instead, the crucial difference is simply that EM and ECM are about the concrete material realisation of certain mental states, and not about how meaning, sense, or content is determined. These two kinds of externalisms need not be in contradiction; they simply deal with completely different matters. (Putnam's account will be discussed in Sect. 2.2.2.)

EM and ECM also differ from *panpsychism* (see, e.g. Strawson 2006) in that panpsychism concerns mental *character*. EM and ECM claim neither that a notebook or a blind person's cane is cognitive or conscious nor that they contain "proto" cognitive/conscious elements. These tools can only become part of the realising material basis for some cognitive/conscious state when properly coupled with an (already cognitive/conscious) agent—and this claim is far from panpsychism. Besides, panpsychism is not empirically testable but remains in the analytic sphere.[21] To conclude, EM and ECM do not have almost anything in common with semantic externalism or panpsychism (except the label 'externalism' in the former case).[22]

1.5 Conclusions and Methodological Remarks

In recent decades, there has been an increase in the amount of literature on naturalistic, empirically responsible philosophy of mind. Rather than relying on speculative a priori thought experiments and pure rational reasoning, naturalistic philosophy of mind is in close interplay with empirical research and experiments in fields such as cognitive sciences, cognitive psychology, neurosciences, psychiatry, and robotics. My research shares this empirically oriented methodology and multidisciplinarity.

I use thought experiments as little as possible. Sometimes they can be useful in illustrating a phenomenon, such as the example of Otto and his notebook. That particular thought experiment is a good example because it doesn't contain any leaps into science fiction but deals with relying on a mundane pair of notebook and pencil. One especially problematic (but

[21] Although some panpsychists might oppose this. For example, *integrated information theory* is said to entail panpsychism, and it is promoted as a scientifically testable theory (see, e.g. Tononi 2008).

[22] Yet another account to be contrasted with is *naïve realism*, according to which the *objects of perception* constitute the experiences (Martin 2004). As Clark (2009, 968) has pointed out, naïve realism is a "more metaphysical" argument than ECM. ECM targets the material vehicles of experiences, not their objects or contents, and is a more empirically oriented position.

common) aspect in philosophy of mind is that people tend to appeal to "future technology" or "future development in neuroscience". My position is not dependent on any "future" inventions but is rooted in existing artefacts, technology, and science.

Not only can the empirical sciences guide philosophy, but philosophy also has a lot to offer the empirical sciences. Furthermore, these topics are related to ethical and societal questions. Apart from being intriguing philosophically, the question about the borders of the mind can also have direct implications on our lives. In democratic societies, citizens' bodies and brains cannot be fiddled with without their permission. However, if the boundaries of our cognitive and experiential functions lie elsewhere, perhaps the legal questions should be adjusted accordingly. Perhaps hacking a laptop should be seen as a personal assault rather than merely an assault on one's property.[23] Skin and skull is not the answer.

References

Bayne, T. (2007): "Conscious States and Conscious Creatures: Explanation in the Scientific Study of Consciousness". *Philosophical Perspectives* 21, 1–22.

Block, N. (2002): "Concepts of Consciousness". In Chalmers (ed.): *Philosophy of Mind, Classical and Contemporary Readings.* New York: Oxford University Press, 2002, 206–218.

Bruno, M. G. (2013): *The Extended Conscious Mind.* University of Arizona. (PhD-dissertation).

Brooks, R. A. (1990): "Elephants don't play chess". *Robotics and Autonomous Systems.* 6(1–2), 139–159.

Carter, J. A., & Palermos, O. (2016): "Is Having Your Computer Compromised a Personal Assault? The Ethics of Extended Cognition". *Journal of the American Philosophical Association*, 2(4), 542–560.

Chalmers, D. (1996): *The Conscious Mind: In Search of a Fundamental Theory.* New York: Oxford University Press.

Chalmers, D. (2008): "Foreword". In Clark, A.: *Supersizing the Mind: Embodiment, Action and Cognitive Extension.* Oxford, UK: Oxford University Press.

Chalmers, D. (2017): "Extended Cognition and Extended Consciousness". In Colombo, M., Irvine, E., & Stapleton, M. (eds.), *Andy Clark and his Critics.* Wiley-Blackwell (forthcoming).

[23] Carter and Palermos (2016) have argued that if we accept the hypothesis of extended cognition, we have to update our notion of *personal assault* to include also intentional harm done towards the gadgets that are properly coupled to the system.

Chalmers, D. (2022): *Reality+: Virtual Worlds and the Problems of Philosophy*. UK: Allen Lane.
Chemero A. (2009): *Radical Embodied Cognitive Science*. Cambridge, MA: MIT Press.
Clark, A. (1997): *Being There: Putting Mind, Body, and World Together Again*. Cambridge, MA: MIT Press.
Clark, A. (2003): *Natural-Born Cyborgs: Minds, Technologies, and the Future of Human Intelligence*. Oxford, UK: Oxford University Press.
Clark, A. (2008): *Supersizing the Mind: Embodiment, Action and Cognitive Extension*. Oxford, UK: Oxford University Press.
Clark, A. (2009): "Spreading the joy? Why the machinery of consciousness is (probably) still in the head". *Mind*, 118(472), 963–993.
Clark, A. (2012): "Dreaming the whole cat: Generative models, predictive processing, and the enactivist conception of perceptual experience". *Mind*, 121(483), 753–771.
Clark, A. (2013): "Whatever next? Predictive brains, situated agents, and the future of cognitive science." *Behavioral and Brain Sciences*, 36 (3), 181–204.
Clark, A. (2015): "Radical predictive processing". *The Southern Journal of Philosophy*, 53, 3–27.
Clark, A. (2016): *Surfing uncertainty: Prediction, Action, and the Embodied Mind*. Oxford: Oxford University Press.
Clark, A. & Chalmers, D. (1998): "The Extended Mind". *Analysis* 58: 1, 7–19.
Colombetti, G. (2014): *The Feeling Body: Affective Science Meets the Enactive Mind*. Cambridge, Massachusetts: The MIT Press.
Colombetti, G. & Thompson, E. (2008): "The Feeling Body: Toward an Enactive Approach to Emotion". In Overton W., Muller U., & Newman J. (eds.): *Developmental Perspectives on Embodiment and Consciousness*. New York: Lawrence Erlbaum, 45–68.
Colombetti, G. & Krueger, J. (2015): "Scaffoldings of the affective mind". *Philosophical Psychology*. https://doi.org/10.1080/09515089.2014.976334
De Jaegher, H., & Di Paolo, E. (2007): "Participatory Sense-Making: An enactive approach to social cognition". *Phenomenology and the Cognitive Sciences*, 6(4), 485–507.
De Jaegher, H., Di Paolo, E. & Gallagher, S. (2010): "Can social interaction constitute social cognition?" *Trends in Cognitive Sciences*, 14(10), 441–447.
Fodor, J. (1975): *The language of thought*. Cambridge, MA: MIT Press.
Gallagher, S. (2005): *How the Body Shapes the Mind*. New York: Oxford University Press.
Gallagher, S. (2008): "Intersubjectivity in perception". *Continental Philosophy Review*, 41: 163–178.
Gallagher, S. (2018): "The Extended Mind: State of the Question". *The Southern Journal of Philosophy*, 56(4): 421–447.

Gibson, J. J. (1979): *The Ecological Approach to Visual Perception*. Boston, Houghton Mifflin.
Haugeland, J. (1998): "Mind embodied and embedded". In: *Having Thought: Essays in the Metaphysics of Mind*. Harvard University Press, Cambridge, 207–240.
Hurley, S. (1998): *Consciousness in Action*. Cambridge, MA: Harvard University Press.
Hurley, S. (2010): "The Varieties of Externalism". In Menary, R. (ed.), *The Extended Mind*. Cambridge, MA: MIT Press, 101–153.
Hutchins, E. (1995): *Cognition in the Wild*. Cambridge, MA: The MIT Press.
Hutto, D. & Myin, E. (2012): *Radicalizing Enactivism: Basic Minds without Content*. Cambridge, MA: The MIT Press.
Hutto, D. & Myin, E. (2017): *Evolving Enactivism: Basic Minds Meet Content*. Cambridge, MA: The MIT Press.
Kirchhoff, M. D. & Kiverstein, J. (2019): *Extended Consciousness and Predictive Processing: A Third Wave View*. London, UK: Routledge.
Kirchhoff, M. D. & Kiverstein, J. (2020): "Attuning to the world: The diachronic constitution of the extended conscious mind". *Frontiers in Psychology*. 11:1966.
Kiverstein, J. (2016): "The Interdependence of Embodied Cognition and Consciousness". *Journal of Consciousness Studies*, 23(5–6): 105–137.
Kiverstein, J. & Clark, A. (2009): "Introduction: Mind Embodied, Embedded, Enacted: One Church or Many?" *Topoi*, 28(1), 1–7.
Kiverstein, J. & Farina, M. (2012): "Do Sensory Substitution Devices Extend the Conscious Mind?" In Paglieri, F. (ed.): *Consciousness in Interaction: The role of the natural and social context in shaping consciousness*, Amsterdam: John Benjamins, 19–40.
Kirsh, D., & Maglio, P. (1994): "On distinguishing epistemic from pragmatic action". *Cognitive Science*, 18, 513–549.
Levine, J. (1983): "Materialism and qualia: The explanatory gap". *Pacific Philosophical Quarterly*, 64, 354–361.
Loughlin, V. (2012): "Sketch this: Extended mind and consciousness extension". *Phenomenology and the Cognitive Sciences*, 12, 41–50.
Martin, M.G.F. (2004): "The Limits of Self-Awareness". *Philosophical Studies*, 120, 37–89.
Menary, R. (2007): *Cognitive Integration: Mind and Cognition Unbounded*. New York: Palgrave Macmillan.
Menary, R. (ed.) (2010a): *The Extended Mind*. Cambridge, MA: MIT Press.
Menary, R. (2010b): "Introduction: The Extended Mind in Focus". In Menary R (ed.) *The Extended Mind*. Cambridge, MA: MIT Press, 1–26.
Menary, R. (2010c): "Cognitive Integration and the Extended Mind". In Menary R (ed.) *The Extended Mind*. Cambridge, MA: MIT Press, 227–244.
Merleau-Ponty, M. (1945/1962): *Phenomenology of Perception*. Trans. Colin Smith 1962. London: Routledge Press.

Nagel, T. (1974): "What is it like to be a bat?" *Philosophical Review* 83, 435–450.

Newen, A., Gallagher, S. & De Bruin, L. (2018): "4E Cognition: Historical Roots, Key Concepts, and Central Issues". In Newen, A., Gallagher, S. & De Bruin, L. (eds.) *The Oxford Handbook of 4E Cognition.* Oxford: Oxford University Press.

Noë, A. (2004): *Action in Perception.* Cambridge, MA: MIT Press.

O'Regan, K. & Noë, A. (2001): "A sensorimotor account of vision and visual consciousness". *Behavioral and Brain Sciences.* 24(5): 939–973.

Pepper, K. (2014): "Do sensorimotor dynamics extend the conscious mind?" *Adaptive Behavior*, 22(2), 99–108.

Phillips, I. (2021): "Blindsight is qualitatively degraded conscious vision". *Psychological Review*, 128(3), 558–584.

Putnam, H. (1975): "The Meaning of 'meaning'". In Gunderson, K. (ed.), *Language, Mind, and Knowledge. Minneapolis*: University of Minnesota Press.

Rowlands, M. (2010): "Consciousness, Broadly Construed." In Menary, R. (ed.) The Extended Mind. Cambridge, MA: MIT Press, 2010, 271–294.

Rupert, R. D. (2009): *Cognitive Systems and the Extended Mind.* Oxford: Oxford University Press.

Silberstein M. & Chemero A. (2011): "Complexity and Extended Phenomenological-Cognitive Systems". *Topics in Cognitive Science*, 4 (1), 35–50.

Silberstein M. & Chemero A. (2015): "Extending Neutral Monism to the Hard Problem". *Journal of Consciousness Studies*, 22 (3–4), 181–194.

Sterelny, K. (2010): "Minds: Extended or scaffolded?" *Phenomenology and the Cognitive Sciences*, 9, 465–481.

Strawson, G. (2006): "Realistic Materialism: Why Physicalism Entails Panpsychism". *Journal of Consciousness Studies*, 13(10–11): 3–31.

Sutton, J. (2010): "Exograms and Interdisciplinarity: History, the Extended Mind, and the Civilizing Process". In Menary, R. (ed.) *The Extended Mind.* Cambridge, MA: MIT Press, 2010, 189–226.

Thelen, E., & Smith, L. (1994): *A dynamical systems approach to the development of cognition and action. Cambridge*, MA: MIT Press.

Thompson, E. (2007): *Mind in Life: Biology, Phenomenology, and the Sciences of Mind.* Cambridge, MA: Harvard University Press.

Thompson, E. & Stapleton, M. (2009): "Making Sense of Sense-Making: Reflections on Enactive and Extended Mind Theories". *Topoi* 28(1): 23–30.

Tononi, G. (2008): "Consciousness as Integrated Information: A Provisional Manifesto". *Biological Bulletin* 215, 216–242.

Uexküll, Jakob von (2010/1934): A Foray into the Worlds of Animals and Humans: With a Theory of Meaning. Translated by Joseph D. O'Neil, Minneapolis/London: University of Minnesota Press.

Varela F., Thompson E., & Rosch E. (1991): *The Embodied Mind: Cognitive Science and Human Experience*. Cambridge, MA: The MIT Press.

Vold, K. (2015): "The parity argument for extended consciousness". *Journal of Consciousness Studies* 22: 16–33.

Ward, D. (2012): "Enjoying the spread: Conscious externalism reconsidered". *Mind*, 121(483), 731–751.

Wheeler, M. (2010): "In defence of extended functionalism". In Menary, R. (ed.) *The Extended Mind*. Cambridge, MA: MIT Press, 245–270.

Wheeler, M. (2015a): "Not What it's Like but Where it's Like: Phenomenal Consciousness, Sensory Substitution, and the Extended Mind". *Journal of Consciousness Studies*, 22(3–4): 129–147.

Wheeler, M. (2015b): "Extended Consciousness: an Interim Report". *The Southern Journal of Philosophy. Spindel Supplement: Alternative Models of the Mind*. Vol. 53, 155–175.

Wilson, R. A. (2004): *Boundaries of the Mind: The Individual in the Fragile Sciences: Cognition*. New York: Cambridge University Press.

CHAPTER 2

Arguments for Extended Conscious Mind

2.1 Introduction

In this chapter, I will give three arguments for ECM. The basic idea behind them is quite simple: if we accept the general 4E-framework, we have no other choice than to accept ECM. Hence, in a broad sense, the most important argument for ECM is simply the whole 4E-framework. For this chapter, I have picked up the most important subtypes of it regarding ECM. The inference relations are as follows[1]: 1) If EM, then ECM. 2) If sensorimotor enactivism, then ECM. 3) If the embodiment thesis, then ECM. They are presented in this order in the following sections.[2]

This alone (showing that ECM follows from the three other theories) is an important result—regardless of whether we accept these three theories or not. It is important because several writers argue for the other three theories but deny ECM.[3] This chapter shows that these views are often

[1] Even though I use the terms 'inference' and 'implication', in none of these arguments the inference relation is a *logical inference* or *material implication*. They are rather a posteriori claims that are to be tested on empirical rather than on logical grounds.

[2] The last remaining E (the embeddedness thesis) makes an exception, because it would not lead to ECM since it is by definition a causal account, whereas EM and ECM are constitutive accounts.

[3] For example, Hutto and Myin (2012, 2017); Colombetti (e.g. 2014); Wilson (2004).

P. Telakivi, *Extending the Extended Mind*, New Directions in Philosophy and Cognitive Science,
https://doi.org/10.1007/978-3-031-35624-7_2

inconsistent—however, not always, for there are at least two exceptions. First, it is consistent to commit oneself to a narrow version of (dispositional) EM but deny ECM. However, a broader reading of EM leads to ECM. Moreover, among the supporters of EM, this broader understanding is very common and widespread: it is what is usually understood by EM at least amongst the philosophers of mind (it is supported, e.g. by Clark). Second, it is possible to accept the embodiment thesis (experience is embodied) but stop there and deny that experience could spread outside the bodily frontiers. However, accepting the embodiment thesis is a slippery slope leading towards ECM, as will be shown in the third argument. However, one of the premise theories, viz. sensorimotor enactivism, differs in this sense: I claim that there is no consistent way to accept it but still deny ECM. All these subtleties and refinements will be clarified and analysed in the following sections.

Thus, there are two ways the reader can receive the arguments in this chapter. First, they can be taken as a clarification of what follows from the three premise theories, without the need to commit oneself to any of them. If understood this way, the results could even be used against the premise theories because a critic of the externalist views could say that if these theories lead to ECM, it is a good enough reason to consider the premise theories false. Even if used in this way, the importance of the matter presented in this chapter remains—whether we are for or against externalism, we have to be consistent in our views. However, my argument for ECM is stronger than this conditional acceptance. Second, the arguments can be taken as reasons why we should accept *both* the premise theories *and* ECM. I claim that the three theories are actually valid/true, therefore so is ECM. Moreover, one does not need to accept all three arguments to reach ECM; even one of them is enough.

2.2 First Argument for ECM: Parity Argument

Most of the proponents of EM have not even considered the possibility of ECM. Clark (2009a, 2012) makes an exception. As said before, he is probably the best-known supporter of EM and at the same time the best-known denier of ECM. However, it is not an easy task to both hold on to EM and deny ECM. Clark needs to use a very complex system to keep the counterbalance, and he ends up making controversial claims in order to uphold EM.[4] Clark's attempt shows that it is very difficult to argue for EM but leave ECM aside. The arguments from straightforward internalists are at least more consistent than Clark's in this case.

This first argument for ECM adopts a treatment opposite to Clark's. It proceeds from EM to ECM. The argument is led by two questions in need of answers: first, why EM? And second, why the bridge from EM to ECM? I will concentrate on the second question since the first question has been discussed comprehensively in the literature for the last 25 years.

For the weaker version of this argument (*if* we accept EM, ECM follows—without taking a stance on whether we should accept EM or not), we don't need any backup for EM, obviously. In the weaker case, only the bridge from EM to ECM needs arguments. However, the stronger version (if we accept EM, ECM follows—and we have good reason to accept EM; therefore we need to accept ECM as well) requires both those "good reasons" for EM, alongside the arguments for the inference relation.

As mentioned before, the literature is full of arguments for (and against) EM. The leading arguments can be found indubitably in Andy Clark's works (e.g. 2007a, b, 2008, 2010a). Besides Clark, Menary (2010a) has edited a worthwhile compilation of defences and criticisms. As mentioned in the previous chapter, the arguments for EM have been classified into first-wave and second-wave arguments (the third wave consists of ECM already, so it is out of this discussion), depending on what aspects they stress (see Menary 2010b, c).

First-wave arguments rely on the *parity principle*, which draws on the functional parity or uniformity of the (putative) external and internal substrates of cognitive processes. As long as the functional role is the same, it doesn't matter whether the apparatus is inside or outside the skin and skull. The parity principle was first and foremost an intuition pump whose purpose was—alongside the Otto example—to question the Cartesian prejudices looming behind the internalist account of the mind.

The second-wave line of argumentation drops the parity requirement (or parity *metaphor*, depending on the interpretation) and instead states that the external processes don't have to functionally mimic any

[4] I will discuss Clark's objections in full detail in Chap. 5. In a nutshell, Clark's argument on high bandwidth (according to which conscious experience requires high-bandwidth processing, and only neural processing is capable of the task) is problematic at least in two ways. First, it is based on dubious assumptions about the body as a low-pass filter that doesn't have empirical support. Second, even if it were true, it would also debunk occurrent EM (which is certainly an unwanted result for Clark).

brain-based cognitive state, but that they complement them instead. Second-way arguments for EM include, for example, *the complementarity argument* (Sutton 2010) and *the cognitive integration argument* (Menary 2007). Instead of the parity principle "the motivation is found in the brute fact of our embodiedness and our bodily manipulation of environmental vehicles" (Menary 2010c, 233).

Behind all these arguments is the simple idea that our borders are not set in stone. As I mentioned in the previous chapter, Hurley (2010) emphasised that there is no reason to assume a "magical" membrane between the brain and its body and environment. The dynamic processes criss-cross between all of them, and they don't respect membranes, magical or mundane. Of course, Hurley's way of putting it is not at odds with the parity (nor complementarity) arguments, but it avoids the problems that can potentially arise from the parity principle. If the parity principle is understood literally instead of as an intuition pump, it can lead to a strong commitment to functionalism and to an impression that extension is only to be detected with the means of comparing it with internal processes as if the internal brain processes were a default case. This criticism towards the parity principle has been articulated by many writers (e.g. Di Paolo 2009; Menary 2006; Wilson 2010). Others have criticised the critique for misunderstanding the aim of the parity principle (e.g. Clark 2007a, b, 2008; Wheeler 2010). Kiverstein and Clark (2009, 3) sum up the defence:

> The parity principle was introduced only to engage our rough sense of what might belong to the domain of the cognitive. It was never a requirement that the external entities perform the same or similar role to processes taking place internal to the individual. Parity demands that we count external elements as forming a part of the machinery of cognition even if their causal contribution is unlike anything we find within the head.

This shows that Clark doesn't commit himself to a dubious interpretation of the parity principle, but still, it doesn't remove the fact that the original formulation of the parity principle by Clark and Chalmers (1998) had all the ingredients to encourage these misinterpretations.

Overall, I think the general 4E-framework that was presented in the previous chapter and the numerous arguments formulated in the literature provide good enough reasons to accept EM. It is not considered an overly controversial claim in philosophy of mind anymore, which is also backed up by the fact that it has been used extensively in other fields, for example, in philosophy of psychiatry, cognitive neurosciences, and art and literature studies.

The following, and more radical step, is to be taken from EM to ECM. At least the following three writers have argued that EM cannot be accepted without accepting ECM alongside it. According to Ken Pepper (2014), if we accept the wider commitments of EM and the lessons from sensorimotor enactivism that Clark needs to accept for his own EM to stay safe, we cannot but accept ECM as well. Pepper discusses ECM from many angles, and I will return to his views in later parts of this book.

Victor Loughlin (2012) argues that some cases of Clark's EM already involve conscious experience, even though Clark didn't intend so (i.e. the example of an artist and her sketchpad). I think Loughlin's argument is on the right track. It touches on the same differentiation that I will present in the following sections, that is, between the occurrent and dispositional EM. However, he fails to explicitly state the difference between the two types and how one leads to ECM and the other doesn't, and that slightly reduces the power of his argument.

Furthermore, Karina Vold (2015) finds Clark's reasons to reject ECM wanting. She pursues the strategy of criticising Clark's criticism based on high bandwidth and claims that the arguments Clark is using for EM work for ECM as well, and Clark hasn't been able to consistently disprove this result. Vold's criticism towards Clark's high-bandwidth argument is very apt, and I fully agree with it (see discussion in Chap. 5). However, in order to argue for ECM, she ends up using a thought experiment, where instead of replacing neurons with silicon chips (as, e.g. in Searle's version), the whole silicon apparatus is placed outside the head and connected to the brain with wires through one's ear. Even though this argument is supposed to give support for ECM, I think it is not the most profitable way to proceed, since it uses the same methods as the opposing camp, viz. brain-in-a-vat scenarios.

Although these three writers' argumentation differ from each other in some aspects, all of them still explicitly state that ECM follows from EM. It should be noted, however, that opposing views have been presented as well. Most importantly, Michael Wheeler (2015a, b) supports EM on functionalist grounds but denies that ECM should follow from it. I will discuss his view later in this chapter.

In principle, yet another way to proceed from EM to ECM would be to claim that all *real* mental states are conscious or at least potentially conscious. Something like this is proposed by John Searle (1992) and Galen

Strawson (1994). Given EM, and given that cognitive states are mental states and therefore conscious states, ECM follows. However, I do not support this view. As presented in the first chapter (1.2), cognition and consciousness are not identical but can be separated at least conceptually. It would not be reasonable to cut the psychological sphere to cover only the phenomenal side.

Of course, this line of argumentation could be used not only as a backup for ECM but also as an objection to EM—if all mental states need to be conscious and the dispositional state described in the Otto example is not conscious, it follows that it is actually not mental. Clark and Chalmers (1998, 16) note this possibility: if dispositional beliefs are not considered real beliefs, because "the only true beliefs are occurrent beliefs", then Otto would not count as an example of EM (and neither would Inga have a real belief before deciding to go to MoMA). Obviously, this is a criticism only against dispositional EM, not against occurrent EM, let alone ECM. (Clark and Chalmers (1998) themselves don't talk about occurrent vs. dispositional EM, and certainly not about ECM.) Nevertheless, as already mentioned, I don't think there are good reasons to accept the claim that all mental states have to be conscious or occurrent, and hence also the dispositional EM survives this criticism.

In what follows, I will present what I think is the most important reason why one cannot consistently support EM and deny ECM. I will approach the matter from two directions. *First*, I appeal to the close ties between cognition and consciousness and defend a view that they overlap in many situations. As was discussed in the previous chapter, the strict separation stems from theoretical traditions that have been predominant approximately in the last half of the twentieth century. The cognitivist paradigm in cognitive sciences (and its influence on analytic philosophy of mind) has deepened the gap. The work in the 4E-field has shown that we should drop both the cognitivist paradigm and the deep-rooted gap between cognition and consciousness.[5] *Second*, I argue that there is a version of EM that includes experiential states, that is, ECM can already be found in one version of EM (that, e.g. Clark himself supports). These two slightly different strategic pathways—whose aim is still to reach the same destination—will be discussed in the following sections.

[5] Instead of the vocabulary I use, some writers speak about *non-conscious cognition* and *conscious cognition* (see, e.g. Kiverstein 2016). Possibly that vocabulary might be more neutral, but for the sake of clarity, I use the terms *cognition* and *consciousness*.

2.2.1 Cognition and Consciousness Are Often Inseparable

Let us start by considering the first option. The first way to proceed is to propose that cognition and consciousness should not be separated in the categorical manner that they have been separated in (especially analytic) philosophy of mind during the last decades. The motivation behind this idea comes from a critical stance towards classical cognitive science (cognitivism), and it has been addressed in some recent enactivist writings (Silberstein and Chemero 2011, 2015; Kiverstein 2016).

As was pointed out in the first chapter, cognitivism has promoted the view that cognition is inherently computational. When understood as computation, it makes sense that cognition has been thought of as independent from consciousness, for it is difficult to see how conscious experience could be explained in computational terms. Therefore, there is a gap separating the two, and that is reflected, for example, in the famous arguments in analytic philosophy of mind such as the *explanatory gap* and *philosophical zombie* arguments.

Let us briefly consider the aftermath of the famous zombie thought experiment (for more details of the argument itself, see, e.g. Chalmers 1996). According to this argument, conceiving of creatures that are functionally identical to us means conceiving identity in terms of *cognitive processes*. The phenomenal side is ruled out because allegedly it is not based on those very same functions—therefore, according to the argument, there is nothing it is like to be a philosophical zombie. When the concepts of cognition and consciousness are separated in such a manner, this argument becomes *conceivable*, a logical possibility. However, this conceivability is only due to assuming that cognition and consciousness can be separated in the first place. Without the separation, the argument is meaningless: zombies would not be conceivable. Creatures functionally identical to us in cognitive terms are also identical to us phenomenologically (see the discussion in Kiverstein 2016, 132).

Consider the two ways to understand the concept of cognition that was presented in the previous chapter (Sect. 1.2). If we conceive cognition in the enactivist way, that is, based on skilled, embodied interaction with the world that already comprises the experiential side, the gap dissolves, and there is no longer a need for principled separation. Kiverstein (2016) draws a picture of cognition and consciousness as entangled and inseparable.[6] Instead of "atomic sensations", that is, qualia that are based on intrinsic properties of sensory qualities (pain is intrinsically unpleasant/painful and the experience of redness is intrinsically red), Kiverstein builds his theory on enactivism and especially on Gibsonian affordances.

Phenomenal experience is based on *everyday human life*—on the practical engagements we have with our surroundings. We primarily experience affordances rather than properties of sensations. Affordances are opportunities (and things to avoid) that the world offers, based on the subject's capabilities and needs. They exist in the relational domain of the subject and her environment. A chair affords sitting for me, a ladder to jump on the table where's food for a dog, and a scaffolding to build a web for a spider. The phenomenality of experiences is explained by our "intentional directness" to the world. What we experience are the "relevant affordances that move us to act in ways that improve our situation in the world" (Kiverstein 2016, 121).

When the notion of phenomenal consciousness is considered this way, it becomes a lot easier to see how cognition and consciousness are not very different from each other. Both are based on the same thing—the agent's skilled interactions in its environment. (*Skilled* is an important aspect here; it refers to the *sense-making* stance that is gained in the interactions.) Cognition is not the computation of mental representations, and consciousness is not something alien to the natural world and explanation. There are no mysterious "qualia". In my experience of a red apple, there is no separable "raw feeling" that would plainly consist of the property of redness itself. Rather, I see an edible, perhaps even a delicious object, and my *knowing* or *believing* that it is a fruit, that I can eat it, and my visual, olfactory, and tactile *perception* of it are not separate from how I experience a red apple.

This kind of understanding of phenomenal consciousness is in particular based on the ideas presented by the *radical enactivists* (Silberstein and Chemero 2011, 2015; Hutto and Myin 2012, 2017), who bridge the gap

[6] Kiverstein also criticises Block's (2002) differentiation between *phenomenal* and *access* consciousness. "Such a distinction only appears necessary because of a computational theory of cognition tailor made for explaining unconscious cognitive processes" (Kiverstein 2016, 109). I am also critical towards a certain use of the differentiation, viz. when the motive is to set phenomenal consciousness (understood as an abstract property of what-it-is-like-ness) in its own sphere, dissociated from report, memory, reasoning, and so on. However, if used more as an epistemic tool, that is, as describing different aspects of experiential–cognitive processes but not aiming to dissociate them, I think the differentiation still has its place.

between cognition and consciousness (or the *mind–mind problem* as they call it). Instead, the experience extends jointly and inseparably with cognition: all mental processing is based on coupled animal–environment systems. They describe the phenomenological–cognitive systems in terms of affordances, niche construction, and evolution.

> [W]e can understand phenomenology and cognition as inseparable and complementary aspects of coupled brain–body–environment systems. On our view, experience is cognition and cognition is experiential. Our cognitive, conscious, and behavioural capacities co-explain and co-determine each other dynamically [...] The systems that cognitive scientists have identified as extended cognitive systems are in fact extended phenomenal-cognitive systems. (Silberstein and Chemero 2011, 40–41)

If we accept this view, what remains of the difference between cognition and consciousness? Are they simply identical, and is the separation only due to a theoretical–philosophical misunderstanding? The radical enactivists don't have a clear stand on this. However, Kiverstein's (2016) suggestion is worthwhile. According to him, we should understand the linkage as something that happens outside of psychological experiments in laboratories and thought experiments in philosophers' writing desks. To put it concisely, *cognition is experiential in everyday-life situations.* They usually occur together. In "normal" situations, "cognitive processes do not unfold independently from conscious processes" (Kiverstein 2016, 114).

However, they are *not identical.* Cognition and consciousness are still separate conceptually, and not only conceptually, but there are occasions when they occur separately. Especially clear are occasions of cognition without consciousness. Psychological experiments have shown that perception can occur without conscious awareness or with degraded conscious awareness, for example, in phenomena such as blindsight, change blindness, and binocular rivalry (see, e.g. O'Regan et al. 1999; Prinz 2015; Rensink et al. 1997). Also dispositional beliefs might be considered occasions of non-conscious cognition (even though some people deny that they count as mental states at all—but this of course depends on how we define mental states). Overall, cognition comprises several different features (e.g. attention, memory, and problem solving, see the list in Sect. 1.2.), and it is clear that not all of them are experiential.

Nevertheless, the fact that cognition and consciousness *can* occur separately doesn't entail that they are separated all the time or in the cases we are interested in. The anomalous perceptual states mentioned above (blindsight, change blindness, and binocular rivalry) are conducted in controlled conditions that are "artificial in so far as they abstract away from the normal ecological contexts in which unconscious and conscious cognitive processes normally occur" (Kiverstein 2016, 114). The cases that I am referring to when discussing the similarities of EM and ECM are exactly the above-described everyday-life mundane tasks, such as sketching with pen and paper, dancing to choreography, or deciding whether an apple is good for eating.

To put it in a nutshell, we don't become Chalmers's zombies when undergoing cognitive tasks. Several everyday cognitive processes (such as recognising an apple as food) are experiential, and experiential states (such as seeing the appleness of an apple) are cognitive. A rigid differentiation does not help to solve the empirical questions that, for example, psychology or cognitive neuroscience deal with. The most important question for the empirically oriented sciences of mind is not *Why is there consciousness at all?* but rather *How does the cognitive-experiential system work?* The question of this book is of course the latter one, with a focus on its subquestion: *How is the cognitive-experiential system materially realised?*

2.2.2 *Occurrent EM*

A slightly different way, which however leads almost to the same result, is to point to the *occurrent* side that is already present in EM. Namely, to start from the fact that the theory of EM itself deals not only with dispositional states (such as Otto and his notebook) but also with occurrent processes (such as calculating with pen and paper). Colombetti and Roberts (2014, 1244) make a significant distinction:

> It is important to distinguish two targets of [EM]: (1) standing, enduring mental states, such as dispositional beliefs; and (2) temporary, fleeting occurrent cognitive processes, such as calculating a complex sum at a certain moment. [EM] is the thesis that, sometimes, the material vehicles that realize both (1) and (2) extend beyond skull and skin, into the world.

This line of proceeding leads to the same conclusion as the previous strategy, but the path of reasoning is slightly different. Instances of occurrent EM already cover the experiential side: they are cognitive(-phenomenological) processes that happen during a certain interval of time, of which the experiential side cannot be carved off (at least in real-life situations, conceptually they might be separable, e.g. in thought experiments). Colombetti and Roberts use an example of calculating to refer to an occurrent cognitive process. I suggest that perhaps an even better example of (2) from the above passage is *calculating with one's fingers*. It is more experiential than "calculating a complex sum", which may be quite abstract (even if completed with pen and paper).

Indeed, the only non-experiential aspect in EM is (1) in the previous quote: standing, dispositional beliefs.[7] Now, Clark most definitely supports occurrent EM and has come up with the most illuminating examples in his publications over the last couple of decades. I think it is fair to say that the most important and interesting instances of EM belong to this group, and it is what makes EM such a striking and influential theory. The only cases that are left to the group of dispositional, non-occurrent EM are based on memory, knowledge, or beliefs such as the Otto thought experiment. In the literature, dispositional and occurrent EM are usually not separated, or if they are, EM is taken as a wider theory that covers both (of which the above quote from Colombetti and Roberts is an example). In general, I follow their example (i.e. when I use the term EM, it covers both sides), but in this chapter, and especially in this first argument, the differences between the two sides play a crucial role.

Of course, this investigation revealed not only that occurrent EM leads to ECM but also that dispositional EM does not lead to ECM. But neither does it rule ECM out; the two theories just cover very different phenomena. The Otto example is not the only instance of dispositional EM, for we can also come up with real-life manifestations. Probably, the most obvious one would be our everyday use of the "Otto-method": how we have off-loaded part of our memory in the everyday equipment and technologies, such as memories from a summer vacation to pictures in albums and computers, appointments in calendars, and several other commonplace pieces of information in shopping lists, diaries, smartphones, and so on. O'Regan

[7] Colombetti and Roberts (2014), however, do not aim to support the line of reasoning that the occurrent side in EM leads to consciousness coming on board. Their paper does not discuss the matter. However, I think it is something that everybody who accepts occurrent EM *should* accept.

already in his 1992 article suggested that the world functions for us as an "outside memory".

It is not common amongst 4E-minded philosophers of mind to support only dispositional but not occurrent EM.[8] It is questionable whether the plain dispositional EM would even count as *active externalism* anymore, as Clark and Chalmers (1998) described it. What matters for Otto is his interaction with the *object of perception* (his notebook and the entry that states the whereabouts of the museum), whereas in occurrent EM (and in ECM), the focus is not on the object or content but rather on the *means of perception* (e.g. using a calculator for solving a maths problem). Hence, extension is obtained in different ways in dispositional and occurrent EM.

Dispositional EM rather resembles *content externalism*, of which Putnam's (1975) *Twin Earth* argument is an example. However, if we follow Clark and Chalmers's conduct, dispositional EM still belongs to *vehicle externalism*, because it is supposed to be about the material basis, and not about the contents of the memories or beliefs in the notebook. Nevertheless, I am not sure whether in this case the difference between content and vehicle is crucial, precisely because of the blurring of active and passive externalisms.

Of course, the act of checking one's calendar or childhood photo album, or Otto checking his notebook, is an *active* deed, whereas the external factors responsible for the beliefs about XYZ or water are passively looming in the distance and in history. Hence, perhaps dispositional EM after all belongs to active externalism, but it is only due to it being contrasted with a very different account, namely Putnam's content externalism. (It is good to remember that Clark and Chalmers (1998) use other examples of EM that belong to the group of occurrent EM, such as the example of playing Tetris.) The dispositional side of EM is the only part of EM that *doesn't* lead to ECM. As I have stressed from the beginning of this book, the Otto example doesn't entail ECM.

Susan Goldin-Meadow's (2001) research gives empirical support for the non-instrumental, constitutive role of moving and gesturing for

[8] On first impression, one would think that supporters of (exclusively) dispositional EM could be found amongst epistemologists who support EM. However, this is not the case, at least not straightforwardly. According to, for example, Duncan Pritchard (2010), not only dispositional but also *occurrent* beliefs extend. Since the target of epistemologists is so different from the targets in cognitive science/philosophy of mind, and the 'occurrence' in the 'occurrent belief' is probably understood very differently, those two disciplines cannot be contrasted without much further investigation (which is not among my aims here).

cognitive tasks, such as solving a maths problem. That moving and gesturing is experiential, in addition to being cognitive. Hence, in occurrent cases of EM, such as calculating a maths task with one's fingers, there are both cognitive and experiential sides at work. Calculating is a cognitive task, whereas the embodied act of using one's fingers is an experiential act. Moving one's fingers when counting provides a clear-cut illustration how the cognitive act is also experiential.

Alternatively, consider calculating with pen and paper or with an electronic calculator. Both are unquestionably cognitive acts. Both are also embodied, experiential acts. The difference from counting with one's fingers is just that there is an extra-bodily tool, so the experientiality is based not only on the body parts but also on the external parts. If we admit those acts of using pen and paper and/or using the calculator counts as instances of EM, they should also be counted as cases of ECM. For it is clear that those acts are both cognitive and experiential, and it is impossible to keep those two aspects apart.

Pleading to the occurrent side in EM can perhaps be seen as a more neutral strategy than the previous inseparability argument (in everyday situations), since it doesn't ask us to change our notion of cognition or consciousness. It aims plainly to point out something that has been there all the time, but perhaps just unnoticed by us. However, I think the two strategies are equally important and that they are just two different routes for approaching the very same result: if cognition extends, conscious experience extends as well.

2.2.3 *Wheeler's Objection*

As I mentioned earlier, it is not agreed unanimously that accepting EM would lead to accepting ECM. On the contrary, the majority of the proponents of EM do not accept ECM. However, few have offered arguments for the alleged non sequitur. Apart from Clark (whose objections will be discussed in Chap. 5), Michael Wheeler (2015a, b) is an exception. Along with Clark, he is one of the main developers of EM, but can also be counted as a critic or at least a brakeman of ECM—although Wheeler's objections are less definitive compared to Clark's. Instead of directly arguing against ECM as Clark has done, Wheeler aims to show that the arguments that have been so far presented for ECM do not yet resolve the case for (or against) ECM. In his two "brakeman" publications from 2015, he scrutinises two arguments for ECM: *the relational character of experience* (that is an aspect of sensorimotor enactivism), and *sensory substitution*, with an emphasis on the latter.[9] (I will discuss these criticisms later in this chapter and in the fourth chapter, respectively).

Wheeler's positive arguments for EM are familiar from the first-wave externalism. He has called his view "extended functionalism" (see, e.g. 2010) alongside Clark (2008). For Wheeler, EM is a functionalist thesis according to which cognitive phenomena can be *multiply realisable*, along the lines of the *parity principle*. And, as everybody knows, phenomenal consciousness is out of reach of functional explanation. However, Wheeler (2010, 265, 2015b, 162) admits that functionalism has the same problem with externalist and internalist accounts of phenomenal consciousness; ECM is no worse off than internalist accounts. Good old functionalism that stresses the causal roles of mental states is at a loss with phenomenal consciousness without there being any difference whether its material basis is inside, outside, or both sides of the head.

Commitment to the functionalist parity principle demands that the potentially cognitive or conscious element can play the functional role in question. Now, it has been thought for decades in analytic philosophy of mind that those functional roles cannot have a grip on consciousness because there is an explanatory (Levine 1983) or ontological (Chalmers 2002) gap between the functional roles and phenomenal experiences. I have already mentioned the zombie thought experiment (Chalmers 1996), and besides that, Wheeler (2015b) also brings up the Chinese nation thought experiment (Block 1978) that considers what would happen if every individual member of the Chinese nation would play the role each neuron plays in the brain—would that system as a whole be conscious? Block's aim was to show that it doesn't, and Wheeler seems to go along with him.

For me, the crux of the problem of Wheeler's account is his commitment to the functionalist picture where these kinds of thought experiments are legitimate. One needs to embrace cognitivism-influenced concepts of cognition and consciousness in order to these examples become clear. (I don't claim that all versions of functionalism necessarily

[9] Of course, in order to get a full picture, the critic should attend to all available arguments. Perhaps we will see new contributions in the future. It is also good to note here that in this book, the phenomenon of sensory substitution is not used as an argument for ECM but rather as a descriptive example of how ECM works. See discussion on sensory substitution in Chap. 4.

are consolidated with those concepts, but at least Wheeler's version is.) It is evident that we cannot attain ECM from this approach to cognition and consciousness. Instead, if we follow the enactivist approach and conceive mind (including cognition and consciousness) as an organism's active engagement and sense-making interaction in its environment, the above-mentioned thought experiments lose their grip.

To make things more concrete, Wheeler (2015b, 159) develops a scenario where Otto needs to look constantly at his notebook in order to keep the address of MoMA in his mind. He speculates that Otto would probably have an "occurrent, conscious belief" during those moments when he checks his book, and hence the information in the notebook would count as "the material vehicle of the content of an intermittent conscious belief" about the whereabouts of the museum. If a passer-by would ask him where he is going, Otto would again have a glimpse at the notebook and the occurrent conscious belief would be (re-)instantiated, and thus the result would be a "conscious belief that is widely realized in Otto's brain, body and notebook".

The lesson from this scenario for Wheeler (2015b, 159) is that it leaves the door open for both externalist and internalist explanations of the machinery of conscious experience. On the one hand, the internalist cannot just insist that the content is in the brain, for fear of begging the question. On the other hand, the externalist can't escape the possibility that during the episodes of looking at the notebook the conscious belief is transferred and stored in his brain, and brain only.

I agree with Wheeler about his conclusion that the Otto scenario doesn't entail ECM—neither the original nor the updated version. I have stressed all the way down the line: the Otto thought experiment doesn't lead to ECM. As I said earlier, the dispositional side of EM doesn't lead to ECM (but neither rules it out). But I also argued that the occurrent side of EM leads to ECM. Thus, why this story about the occurrent side of Otto's belief doesn't back up that claim?

It is quite self-evident that there is some kind of consciousness involved when Otto reads the address from his notebook (we can call it awareness or access consciousness). However, that awareness is first and foremost "consciousness of knowledge-that", which is just another way to phrase "occurrent cognitive state" or, to use Wheeler's own words, "conscious belief". That is not the kind of perceptual phenomenal conscious experience ECM is after.

I think it doesn't make sense to try to convert Otto's cognitive action into experiential terms at all because the thought experiment is tailor made for such a different purpose. However, the drawback only applies because the Otto case is a highly speculative thought experiment, and in that context cognition and consciousness can be separated neatly. Instead, if we step out of the laboratory settings and thought experiment worlds, and adopt the view I presented earlier, according to which cognition is experiential in everyday-life situations, also cognitive actions are (often) accompanied by phenomenal consciousness.

Experiential side cannot be carved off from many occurrent cognitive processes—and this is not undermined by the observation that it can be carved off from narrow thought experiments (such as Otto's). Consider the examples I gave about calculating with one's fingers, pen and paper, or a calculator. Or an architect sketching a floor plan for a house. Or a taxi driver using a navigator to orient herself in a city. These are all occurrent cognitive processes from which the experiential side cannot be carved off without the whole status of that process changing completely.

2.2.4 *Summing Up the First Argument*

The two strategies presented are just two different routes for approaching the very same result: if cognition extends, conscious experience extends as well. The core reason why EM cannot be accepted without ECM became apparent from both strategies: cognition and experience cannot be separated in the way they have been separated by the leading supporters of EM. The narrow version of EM (dispositional EM) doesn't lead to ECM, but the wider (occurrent) does—and the latter includes the most important and interesting instances of EM.

A further reason why EM and ECM stand or fall together is that if Clark's counter-arguments against ECM were true, they would also refute his own EM. Especially, his criticism against ECM based on high bandwidth (Chalmers 2008; Clark 2009a, 2012), were it a valid argument, would also—unintentionally—bring down EM. I will argue in Chap. 5 that the high-bandwidth argument doesn't work in general, and hence ECM (and EM) survive. Nevertheless, despite its faults, it is a nice example of how EM cannot be defended independently of ECM. (Needless to say, this is quite the opposite way to interpret the matter from Clark's.) Thus, Chap. 5 provides yet another argument for the claim that ECM follows from EM.

Finally, I discussed Michael Wheeler's objection to ECM based on functionalist reading of EM. He has argued against the very core of this first argument: according to him, EM doesn't lead to ECM. I showed that due to his conceptual toolkit, this conclusion indeed is inevitable, but as I argued earlier in this chapter, this kind of cognitivism-influenced conceptual toolkit is not something that we should adopt. In the end, mine and Wheeler's accounts are not completely at odds, since I can agree with his statement that extended functionalism (understood in his terms) doesn't lead to ECM. Something else is needed, but fortunately we have options for that "something else": it will be dealt with in the rest of this chapter.

2.3 Second Argument for ECM: Sensorimotor Enactivism

Whereas the defenders of EM usually deny ECM, the supporters of ECM are often found amongst sensorimotor enactivists. According to my second argument, if we accept sensorimotor enactivism, ECM follows. Not only do I argue for this entailment, but I also claim that sensorimotor enactivism is an explanatorily powerful theory of perceptual experience, and we should accept it.[10] The benefits of the theory are independent of my purposes with ECM. It provides explanations of some features of perceptual experience that remain unresolved or mysterious in representationalist and cognitivist theories, for example, experiential differences between sensory modalities, and within one sensory modality. That is, sensorimotor enactivism answers the *comparative explanatory gap*. It has been suggested in the literature (Hurley and Noë 2003; O'Regan 2016) that the explanatory gap (Levine 1983), which is a close relative to the hard problem of consciousness (Chalmers 1996), is not the only (putative) gap in explanation. Apart from this *absolute explanatory gap*, there is the comparative gap: the puzzle why different sensory experiences differ in the way they do. Providing an answer to the comparative gap can be considered to be the most important outcome of sensorimotor enactivism.

[10] I use sensorimotor 'enactivism', 'approach', and 'view' synonymously. Proponents include O'Regan and Noë (2001); Hurley and Noë (2003), Noë (2004, 2012), Silverman (2013), Hurley (1998), and Kiverstein (2010). Especially vision is often used as an example of the extended perceptual mode of experience, for example, in O'Regan and Noë (2001), Bruno (2013), Loughlin (2012), and Pepper (2014).

Even though sensorimotor enactivism as a theory in cognitive science is independent of ECM, I argue that a defender of sensorimotor enactivism cannot consistently deny ECM. This argument consists of two steps. First, I will discuss the reasons why we should accept sensorimotor enactivism, and second, the reasons why sensorimotor enactivism leads to ECM. As previously mentioned, I claim that the second step follows quite straightforwardly, even though not all sensorimotor enactivists explicitly address it, and furthermore, Clark (2009a) explicitly rejects the implication (challenges to the implication will be discussed in 2.3.1.).[11] First, I will outline what sensorimotor enactivism actually claims, and then I will sum up what kind of arguments for ECM can be drawn from it.

2.3.1 *Landscapes of Sensorimotor Enactivism*

Understanding perception as an active exploration of the world through sensorimotor skills and patterns is a feature that is very typical of enactivism in general. According to enactivism, the material basis of perception is in the interactions between brain, body, and environment, and consciousness is not considered an epiphenomenal by-product, but it already belongs to cognition (see, e.g. Colombetti and Thompson 2008; Gangopadhyay and Kiverstein 2009).

As explained in the first chapter, enactivism is composed of three main branches: *autopoietic* (e.g. Varela et al. 1991; Thompson 2007), *sensorimotor* (e.g. Hurley 1998, 2010[12]; O'Regan and Noë 2001; Noë 2004, 2006a, 2012), and *radical* (e.g. Chemero 2009; Hutto and Myin 2012, 2017).[13] Even though in this argument I only discuss the second (the sensorimotor branch), this does not mean that the two other branches

[11] Apart from denying the implication (which is the main focus in his 2009 paper), Clark has also criticised "strong sensorimotor models" (see, e.g., Clark 2008, Chap. 8), which includes Noë's account.

[12] Susan Hurley's view is sometimes separated from the sensorimotor branch (e.g., in Gangopadhyay and Kiverstein 2009), since Hurley firmly denies the need for a higher-level integration at the conceptual level (as opposed to, e.g. O'Regan and Noë). Hurley's view can be considered a version of *strong enactivism* (which will be discussed later in this section): perception truly requires action, and not only *knowledge* of the sensorimotor effects of actions. However, the difference found between Hurley and the other sensorimotor enactivists is not essential for the current purposes of settling the premise theory that entails ECM. Thus, I treat Hurley as one of the sensorimotor enactivists.

[13] When I use the term 'enactivism' without any specifications, I refer to the general view that covers the sub-branches.

would be in conflict with it. The types of enactivism do overlap. However, *autopoietic* and *radical* accounts often bypass conscious experience and instead discuss the minimal conditions of cognition, the "lower levels of mind", that is, the cognitive system as an autopoietic system (Varela et al. 1991) or "basic minds" (Hutto and Myin 2012) that is already present in simple life forms, for example, in bacteria. The sensorimotor branch is the one that most directly addresses the issues related to perceptual experience—but it is an independent theory from ECM. Let us now take a closer look at what sensorimotor enactivism involves.

The central feature of the theory lies in the concept of *sensorimotor contingencies* (also referred to as sensorimotor *dependencies*, *loops*, or *patterns*) that are "regularities governing how sensory stimulation depends on the activity of the perceiver" (Seth et al. 2016, 267). They highlight the deep interdependence of the sensory input and motor output—the bodily interactive skilful encounters we have with our environment. Our experiences are tied to sensorimotor contingencies or laws that are at work when we are in bodily interaction with our environment. Perceivers "know how to exploit them to explore and negotiate their environments" (Hurley and Noë 2003, 146). They are said to determine the *content* and *character*[14] of experience (O'Regan and Noë 2001; Noë 2004, 2012; Ward et al. 2017). "Perceptual experience *just is* a mode of skillful exploration of the world" (Noë 2004, 194).

According to the sensorimotor approach, skilful bodily engagements with the environment determine perceptual experiences, and those engagements themselves are constitutive for the content and character of perceptual experience. Importantly, they not only *enable* or *affect*, but do nothing less than *constitute* the experience. Hutto and Myin (2012, 11) write that perceptual experience is "literally constituted by, and to be understood in terms of, concrete patterns of environmental situated organismic activity, nothing more or less". Ward (2012, 738) is no less clear: "the enactivist's claim is that an evolving perceptual relationship

[14] 'Content' here means the *character* of the content, that is, *what* is experienced (compared with *how* the experience is produced—with which vehicles). 'Content' is used in a weak/mild sense, and it doesn't have any technical meaning that would greatly differ from 'character'. It certainly doesn't refer to representations but rather just simply notifies that experiences are about something. See Hurley (2010) and my discussion about the *what*- and *how*-explanations later in this chapter. From now on, instead of referring to both character and content, I sometimes simply refer to character.

with the world, implicating sensorimotor understanding, *is* our experience of the world".

The connection to Gibson's (1979) *affordances* is unmistakable. The subject perceives the affordances in its environment—what it can, for example, grasp, stroke, eat, or avoid. Another source of influence can be traced back to Dewey's *transactionalism* (Dewey and Bentley 1949) that stresses the inseparability of the organism and its environment: the subject's actions, behaviour, and mental states (including higher-order epistemic states, "knowings") cannot be explained without referring to the full organism–environment situation. Similarly, the sensorimotor approach takes into account the whole organism–environment sensorimotor loop.

A similar kind of reasoning can be found in the work of the robotics pioneer Rodney Brooks (1990), who started building robots based on the ideas similar to the sensorimotor approach. Rather than trying to create internal representations or symbols, his robots learn from their sensorimotor actions in their environment by "using the world as its own best model". This case from robotics is interesting because with robots, no doubt is left that there would be some hidden symbol manipulation at play. We know (or at least Rodney Brooks knows) for a fact how these robots function because they are built by humans.[15] In theory, an internalist about the human mind could always appeal to the current incompleteness of neurosciences: perhaps the places for symbols in the brain will be found in the "future", when the neurosciences develop.[16]

2.3.2 *Bridging the Comparative Explanatory Gap*

Sensorimotor enactivism deals with the experiential (i.e. qualitative character) side of perceptual experiences. Some proponents also discuss whether the character determines the vehicle—whether the vehicular side

[15] Robotics offer an interesting case for 4E-theories; they can learn from each other. However, my thesis is about the human mind—and even though I think that the investigation of the possibility (or impossibility) of an artificial mind could also shed light on the human mind—it is, however, not possible to handle these issues here. The questions are so complicated that they are worth their own research.

[16] These kinds of arguments that deal with their problems by appealing to future sciences could, in my opinion, be left to future science, and as we only have access to the present and past science, we would never need to see them again.

is also extended.[17] However, the material basis of experience is not the main explanandum for sensorimotor enactivism (as is it for ECM). It is rather a theory that aims to explain the qualitative side of conscious experiences: qualitative differences between modalities (e.g. visual and auditory) and within one modality (e.g. between different colours or shapes). The sensorimotor approach has an indisputable asset when compared with its internalist competitors: it promises genuinely to *explain* the differences (especially with the notion of *sensorimotor contingencies*), whereas accounts that seek for neural correlates of consciousness—even if they did find them—would always come up with an explanatory gap (O'Regan and Noë 2001, 940–941; Silverman 2013, 152).

For example, in visual exploration, the colour and brightness reflected from objects change in regular, lawful ways when we move around, or the light source or the object moves (O'Regan and Noë 2001, 942). Sensorimotor enactivists stress that to be a visual perceiver is "to be capable of exercising mastery of vision-related rules of sensorimotor contingency" (O'Regan and Noë 2001, 943):

> We have expectations how perceptual experience is likely to change due to our own movement and movement of objects that is independent of us. A first law distinguishing visual percepts from perception in other modalities is the fact that when the eyes rotate, the sensory stimulation on the retina shifts and distorts in a very particular way, determined by the size of the eye movement, the spherical shape of the retina, and the nature of the ocular optics. In particular, as the eye moves, contours shift and the curvature of lines changes. (O'Regan and Noë 2001, 941)

For example, we have an implicit sensorimotor understanding how red objects behave in different lighting. O'Regan and Noë (2001, 951) write that "'red' is knowing the structure of the changes that 'red' causes". The parts that are facing the light look brighter than those that are in the shadow. To know what red is, is to know how it would look different when "colour critical conditions" vary—the colour is "a pattern or structure in appearance" (Noë 2016, 73). The phenomenon of *colour constancy* speaks for this kind of implicit sensorimotor understanding that we possess: despite apparent changes in colour due to change in the illumination

[17] I will get back to the issue why I think it is legitimate to claim that the experiential level determines the material level.

conditions or in the contrasting objects, we still perceive a uniform colour and not a variable appearance (Noë 2004, 127).

This goes for other perceptual experiences as well, such as tactile experience.[18] *Touching* is a good example how the sensorimotor view explains the qualitative differences within a sensory modality. Tactile modality is linked to our movement, to our action, and "to our implicit understanding of the relevant tactile-motor dependencies governing our interaction with objects" (Noë 2004, 98). Touching is not possible without sensorimotor action; we need to sense the resistance of the object to have a tactile experience (see O'Shaughnessy 2000; and Chap. 3).

> For example, the quality of softness of a sponge is constituted by – it is nothing more or less than – all the things that one can do with the sponge when one is checking for softness, and all the ways the sponge reacts when one handles it in this softness-checking way. Anything one can potentially say about the quality of softness of the sponge must ultimately boil down to something about how one potentially can interact with the sponge when one is checking whether it is soft. (O'Regan 2016, 41)

In fact, according to sensorimotor enactivism, not only visual and tactile, but all sensory modalities are constituted by patterns of sensorimotor contingencies. Sensorimotor contingencies also explain phenomenological differences between different sensory modalities. Different sensory modalities are governed by different sensorimotor contingencies: visual experiences are partly governed by eye movements and saccadic movements in a law-like way, whereas for auditory experiences these factors are irrelevant. The sensorimotor contingencies determine why seeing has a certain phenomenological character and hearing has a different character. We as perceivers are familiar with these law-like contingencies, and we know how to exploit them. This is the core reason why sensorimotor contingencies have more explanatory power than neural correlates.

[18] Noë (2016) explains *sensory experiences* such as sensations of pain in a similar manner. Pains are context dependent (e.g., compare having a tattoo vs. a tiger mauling your arm), and as such they contain an accessibility or potentiality that resembles the way colours are accessible to us. However, notice that a sensation of pain is not a perceptual experience, and hence not amongst my examples. Noë makes an exception because usually sensorimotor enactivists only talk about strictly perceptual experiences and exclude "experiences of thought, emotion, and feelings like hunger and pain" (Degenaar and O'Regan 2015, footnote 1).

With neural correlates there is always the accompanying (comparative) explanatory gap without an answer: why is a certain brain chemistry and physics correlated with exactly the experience it is correlated with and not with something else, for example, the experience of seeing rather than hearing? O'Regan (2011, 115) states that "there is no natural way of making such a link". However, this kind of comparative gap doesn't arise with sensorimotor contingencies. Actions themselves serve as an explanation (Hurley and Noë 2003, 146–147). Solving the comparative gap can be counted as a reason why we should embrace sensorimotor enactivism.

2.3.3 *Sensorimotor Reductionism or Sensorimotor Integrationism?*

The character of perceptual experience (such as seeing a round-shaped red object) is given by the *mastery* of sensorimotor contingencies. So the crux is in our *mastery*, *knowledge*, or *understanding* of subject–environment dependencies (O'Regan and Noë 2001, 944; Ward et al. 2017, 371). The word 'knowledge' should not be understood as a propositional attitude but rather as a basic embodied, implicit, practical understanding of the ways our actions affect the world and our perceptions of it. Especially Noë stresses the role of *knowledge*, whereas radical enactivism on the other hand criticises views where its role is too significant. In order to avoid controversial terminology, 'sensorimotor empathy', 'feeling-into' (Chemero 2016), and 'sensing-in' (Colombetti and Torrance 2009) have, among other terms, been suggested.[19]

Besides terminological ambiguities, the above considerations are connected to a deeper tension within sensorimotor enactivists. On the one hand, the theory seems to depend upon concrete bodily action, and on the other hand, mere know-how or understanding of the sensorimotor dependencies seems to be enough. Aizawa (2010a) has named the former "strong enactivism" and the latter "weak enactivism". Are they really two different versions of sensorimotor enactivism, and if so, is the differentiation important considering ECM?

[19] The nature of the sensorimotor contingencies is not a settled issue amongst the enactivists, and the scientific implementations (e.g., in cognitive sciences and robotics) have varied from non-representationalist dynamical interpretations to more representationalist and cognitivist approaches (for an overview see Ward 2012, 371).

Undoubtedly, Clark (2009a, 974) is right to note that "mere potentialities of experience are surely not what is at issue between the friends and foes of ECM". Of course, ECM is talking about actual, not potential experience. However, I do not think we can cut out the sensorimotor tracking abilities from the actions themselves—at least when talking about perceptual experience in humans. The gist of sensorimotor enactivism lies in the inseparability of sensorimotor actions and understanding their effects. What sensorimotor enactivism means by those actions already presupposes and depends on familiarity, attunement, understanding (or "knowledge", if you want) of sensorimotor contingencies.[20] But that "knowledge" is *knowledge-how*: practical, hands on, non-propositional, and embodied in nature.

Keeping Aizawa's differentiation as well as my criticism of it in mind, we might still want to ask: are these sensorimotor actions and the following understanding of the law-like contingencies *everything* that is needed for there to be conscious experience? Is it a necessary and sufficient constitutive basis for perceptual experience?

Of course, it depends on how we understand this "knowledge-how" or "understanding". We can postulate a behaviourism-inspired view where actions are all that matters. Noë (2016) calls *sensorimotor reductionism* a view according to which someone or something who manifested a "sensorimotor profile" typical for some experience would count as having that experience. For example, a robot that acts as if it is in pain would be in pain. Noë criticises this view, but he doesn't directly name any supporters of it. The closest are, according to him, Hutto and Myin (2012) (and probably radical enactivism in general, even though Noë does not say so).

Noë classifies himself as a *sensorimotor integrationalist* (as opposed to a reductionist). The difference with SM-reductionism is that according to Noë's SM-integrationism "we can't give an entirely non-mentalistic account of the substrates of consciousness" (Noë 2016, 71, footnote 3). The non-reducible features in question concern how we achieve the

[20] Of course, we can have a system that is capable of sensorimotor actions but incapable of understanding them in any stronger meaning of "understanding". Consider, for example, a robot vacuum cleaner. Its actions are based on sensorimotor information gained from its actions in the environment. Still, we would probably say it doesn't have the right sort of understanding of the sensorimotor contingencies. However, sensorimotor enactivism is not a theory about vacuum cleaners.

world's presence through our skilful access. He arranges skills of access in three. First, there is our *knowledge* of what and how we can access the world, including conceptual, practical, social, and sensorimotor skills or techniques. Second, there is access to the "real world": it is not the world of sense-data, but neither is it the world of physics, because before physics, "there are tables, chairs, doors". I think he refers to our general situatedness in the world, our attitudes towards it that are based on the *affordances* we can achieve. And third, there is our "concern, caring and interest" attitude towards our environment: we reach out because we care about what there is (Noë 2016, 75–76).

When understood as Noë describes it, SM-reductionism is not a feasible account. I agree with Noë that "something else" is needed, and I think that something else is more or less the same features Noë himself called for. I wouldn't place such stress on *knowledge* how we can access the world, but rather on our primordial situatedness and *sense-making* attitude (cf. Noë's "concern, caring and interest"). The same phenomenon has also been referred to by enactivists as primordial affectivity (Colombetti 2014), affective scaffolding (Colombetti and Krueger 2015; Colombetti and Roberts 2014), and affective framing (Maiese 2015). All these notions stress that the world is not indifferent to us, but it is affectively loaded. Sense-making is needed for there to be conscious experience, such as visual experience of red or a sensation of pain. For us, humans, this caring attitude is present in our sensorimotor actions and contingencies, and that makes sensorimotor reductionism an empty concept, and perhaps a strawman too, if it doesn't have any supporters among sensorimotor enactivists. I think it could potentially be a useful concept when discussing AI systems or robots, but they are out of scope of this study.

To recap, I do not think *sense-making*, *caring about*, or *having an interest in* the outcomes of one's sensorimotor actions should be classified as something "extra". Our sensorimotor actions and an understanding of the contingencies (the right kind of sensorimotor profile) already comprise the sense-making attitude. Thus, sensorimotor reductionism is not a possibility for human minds. For the same reason, strong enactivism also already comprises the know-how of sensorimotor dependencies that weak enactivism promotes. However, an interesting question remains whether there could be weak enactivism without strong enactivism, and this will be discussed in relation to dream experiences in Chap. 5.

2.3.4 Temporal Nature, Virtual Presence

Another important feature is that sensorimotor enactivism is committed to *temporal extension* of experiences, which means that "experience is a mode of temporally extended skilful interaction with the world" (Ward 2012, 737). Perceptual experiences unfold over time, and they are irreducible to any atom-like atemporal states (as opposed to some twentieth-century philosophy of mind views that stressed representations and qualia). To explain consistently temporally extended phenomenology, we need a temporally extended material basis. This means that we need to commit ourselves to an *extensional* model of perceptual experience. It is a view that holds that experiences are dynamic temporal phenomena—all the way from the material basis to the experiential level.[21]

> An extensionalist sensorimotor account, in the first place, explains temporal experience by positing that perceptual awareness of a temporally extended event supervenes on a temporally extended process of interaction between the perceiver and the environment, in such a way that the content temporally tracks the vehicle. (Silverman 2013, 156)

The temporally extended nature of experiences is related to another central feature of the sensorimotor approach. Our relationship with the world is based on evolving (temporal) processes that are explained by the thesis of *virtual presence* or *presence in absence* (Noë 2004, 2006a, 2008). This is the solution of sensorimotor enactivism to some prevalent puzzles of perceptual experience.

We seem to perceive a lot that we don't actually perceive, and we also perceive a lot that we don't notice perceiving. We seem to have a richly detailed, colourful visual experience of the scenery we are facing. However, there is a vast amount of empirical evidence that we (or our brain) are actually not capable of perceiving all that we seem to perceive. This research is well established and accepted by the research community—the possible ambiguities lie in how to interpret these results. Empirical research about blind spots, impoverished peripheral vision (including peripheral colour vision), saccadic smear, change blindness, inattentional blindness, and so on, shows that our visual perception is an impoverished machinery

[21] In the next chapter, I will explain in more detail why ECM and enactivism are forms of the *extensional* model and compare it with *retentional* and *cinematic* models. The temporal and dynamical nature of experiences in general will be discussed in more detail there.

for experiencing the world around us (O'Regan and Noë 2001; Stafford and Webb 2005).

Yet, we don't notice the scarcity. The perceptual experience of the scenery is not monochromatic and blurry in peripheral or parafoveal areas. We are sometimes unaware of dramatic changes in the scenery, as was shown in the famous "invisible gorilla test" (Simons and Chabris 1999) that shows that when we are concentrating on one aspect in our visual field, we easily fail to notice other objects, no matter how striking they are, such as a person walking in and out in a gorilla suit (inattentional blindness). We also tend to fail to notice large changes in our visual field, such as objects disappearing or changing colour—we are unable to track the difference in a picture before and after, or when the picture flickers on and off (change blindness) (see O'Regan et al. 1999). In addition, "most people are completely unaware of how poor their acuity and their color perception are in peripheral vision" (O'Regan and Noë 2001, 951).

This is the puzzle: why do we seem to have a rich visual experience coated with lots of details and colours around us? A common internalist and representationalist response is that the rich visual image is due to our internal representations. We have an inner "picture" of the scenery: the richness of details comes from inner representations being activated. A related response is that the apparent richness is only *illusory* (and hence this view is sometimes called the *grand illusion*). According to this view, we are mistaken in our experience: we *think* we enjoy rich perceptual experiences, but actually we don't (see discussion in Clark 2009a, 973).

The sensorimotor approach denies both the *grand illusion* and the *internal representations* explanations. The richness of colours and details is true for us on the experiential, phenomenological level because we know how to attend to them when needed. In other words, sensorimotor enactivists explain the tension by pleading to sensorimotor contingencies. The details of the scenery are present virtually because of our dynamic sensorimotor access to them. "The flick of the eye, the turn of the head, the movement of the body, brings us the detail we need as we need it" (Noë 2006a, 420). The richness is not based on representations constructed in the brain but rather on our skill-based access to them in the actual environment. Hence, our perceptual experiences are not only determined by neural states but also by actions and the environment.

Sensorimotor enactivism explains *all* perceptual experiences in this manner (even though the focus is often on visual modality, especially in Kevin O'Regan's work). Apart from the above-discussed neurological deficiencies, the theory also explains the more mundane phenomenon that needs no laboratory experiments: the fact that we only see the parts of the objects we are facing and that are visible to us but never the hidden, occluded sides. The retinal image can only offer a view of the *fronts* of the objects; it is impossible to see all the sides at once. However, this is not how objects are experienced. Instead, we experience objects as wholes. We have visual experiences of complete three-dimensional cats, dogs, and tomatoes, not only their unoccluded parts. "When you see a tomato, you only see, strictly speaking, the visible face of the tomato; but it is also true that you are visually aware of the presence of the parts of the tomato which you don't actually see" (Noë 2006a, 414). The same applies for apparent size and shape. We see two trees in our visual field as being the same size, even though the nearer tree looks bigger; and the plate in front of me looks both circular and elliptical. Another example is the already mentioned colour constancy: the wall looks uniformly white, and yet it also looks grey, yellow, darker, lighter, and so on (Noë 2004, 2006a).

In the same manner as earlier, the hidden sides and perceptual ambiguities are explained by sensorimotor contingencies. The world is present to us virtually: with the means of our sensorimotor contact to it, and our understanding of the ways in which our movements and exploratory habits bring the occluded sides to our awareness. We have *access* to the occluded, hidden sides by means of our probing, moving, and acting in the environment. We have learned to anticipate how a shape or colour changes when observed from different angles. This is what Noë means by virtual presence.

Noë's account of virtual presence has been criticised (e.g. Clark 2009a) for putting too much weight on the *potentiality* of experience. If the rich, detailed perceptual (character of) content is present to us only potentially, what does it tell us about the actual experiential contents, which are surely our focus of interest? However, this criticism is based on a mistaken reading of Noë. Ward clarifies Noë's intentions with his notion of virtual presence:

> We should not interpret his remarks on presence as entailing that the detail of the world, or the occluded parts of objects, are present to us in experience 'only potentially' or in some impoverished sense. Whole objects and detailed scenes are embraced in experience, despite facts about the limits of sensory uptake that might appear to make this problematic. (Ward 2012, 737)

Hence, the hidden parts are *not* present to us potentially, but on the experiential level, they really *are* present. Ward points out that Clark assumes that Noë is talking about the sub-personal level of experience. But instead, the notion of virtual presence is supposed to capture how we perceive the world in a phenomenological sense (I will discuss the personal vs. sub-personal question in more details in Sect. 2.3.6.). The virtual presence thesis provides a conception of experience (at a phenomenological, personal level) as an interactive relationship between subjects and their environment—and this conception explains all the above-mentioned deficiencies in our perceptual capacities.

The main features of sensorimotor enactivism can be summarised as follows: 1) Perceptual experience depends constitutively on skilful bodily interactions with the environment and the perceiver's implicit understanding of those dependencies (sensorimotor contingencies). 2) Phenomenal qualities are differentiated in line with the subject's actions, expectations, and understanding of how her movements will change the qualities. 3) Experiences are temporally extended dynamic phenomena. 4) Perception is virtual in the sense that even though we never have a retinal image of a complete scenery, object, colour, or other properties, the details are still present due to their accessibility through our skilful sensorimotor actions.

2.3.5 Arguments Drawn from Sensorimotor Enactivism in Support of ECM

In this section, I sum up what I think are the most central reasons why sensorimotor enactivism leads to ECM, viz., how the features described in the previous section entail ECM. However, let us first take a look at what others have said about this. In his critical paper, Clark (2009a) names the features of sensorimotor enactivism that according to him are the best candidates that support ECM. (His aim is to show that those features *don't* entail ECM.) Since Clark's influential paper, the features he picked up have been, to a certain extent, accepted to capture the most important arguments for ECM (they are referred in e.g. Ward 2012; Clark 2013; Pepper 2014; Kirchhoff and Kiverstein 2019). I agree that they are important arguments, but I don't think they are the only reasons to accept ECM: apart from them, my first and third arguments in this chapter should also be taken into account. In what follows, I review the standard line of arguments from sensorimotor enactivism in favour of ECM and discuss whether they work or not.

Clark (2009a) calls them arguments from 1) variable neural correlates, 2) virtual representations, and 3) dynamic entanglement + unique temporal signature (DEUTS for short). First, ***the variable neural correlates*** argument is based on empirical facts about how changing sensorimotor contingencies affect the brain by means of neural plasticity. Sensorimotor enactivists (e.g. Hurley and Noë 2003) give an extensive overview of empirical studies on neural plasticity that show that neural correlates are not fixed.

An example where the neural processing varies, but the experience remains the same is inverting goggles (e.g. left–right reversing goggles). When the goggles are first used, visual experience is disrupted dramatically; there is conflict between proprioceptive and visual modalities. Sensorimotor contingencies no longer hold, since objects on the left side are not accessible by visually leftward movements. However, after enough practice and use, the distorting effects of the goggles will disappear, and the user will perceive the world normally again. (For the left–right reversing goggles experiment, see Taylor 1962; and the discussion in Hurley and Noë 2003.)[22]

The aim of the variable neural correlates argument is to show that the explanation of experience based solely on neural factors is not sufficient. Instead, experience requires an explanation of the modes of interaction with the environment (i.e. sensorimotor contingencies). The case of inverting goggles shows that the users must learn new ways how sensorimotor contingencies work, and only after the attunement can they have normal experiences again. Thus, it is the characteristic interaction relation that fixes[23] the qualitative character of experience. "[W]hat determines phenomenology is not neural activity set up by stimulation as such, but the way the neural activity is embedded in sensorimotor dynamic" (Noë 2004, 227).

[22] Another example of neural plasticity that is often used is neural rewiring (visual to auditory) in newborn ferrets. Sur et al. (1999) managed to reroute visual input to auditory areas so that the ferrets' visual modality was connected with the brain parts that were usually associated with hearing (discussed in Noë 2004; Hurley and Noë 2003). However, this example is more controversial than the use of goggles or sensory substitution devices due to reportability issues: we cannot ask the ferrets about their perceptual experiences.

[23] Here, 'fixing' and 'determining' means fixing and determining *constitutively*. Technically, the phrase could also be: "The characteristic interaction relation *is* the qualitative character of experience".

The variable neural correlates argument stresses the plasticity of the neural mechanisms, the plasticity of the sensorimotor contingencies, and how they affect each other. However, I don't think it is the strongest argument that sensorimotor enactivism has to offer in support of ECM. Instead, it is best taken as a reminder of prevalent plasticity, considering neural functions and especially perceptual plasticity. It can be read as a version of the well-known *multiple realisability* argument; and in this general sense, it encompasses something essential to externalist views, something that is necessary but not sufficient for them.

Second, the argument from *virtual representations* draws from the same phenomenon of *virtual presence* (or *presence in absence*) that was already introduced in the previous section. It is indeed one of the most central arguments for the role of sensorimotor contingencies, that is, for the role of agent–environment loops in constituting our perceptual experiences. This leads directly to ECM: experience is based on those interactive relationships that contain external parts. (I will discuss below and in the next section why the personal level determines the sub-personal level.)

The third and last argument formulated by Clark (playing devil's advocate), the DEUTS argument, consists of two parts. The first part, the *dynamical entanglement* thesis, holds that the material basis of perceptual experience is a dynamic mess criss-crossing the brain, the extra-neural body, and the environment (see Hurley 1998, 2010). It is impossible to separate the non-neural elements from the brain functions—organisms are very deeply and complexly intertwined systems (Cosmelli and Thompson 2010). Sometimes referred to as *continuous reciprocal causation* (e.g. Clark 2008, 24), these dynamic complex loops are the material basis for perceptual experiences.

Thus, the dynamical entanglement thesis bridges the other features of the sensorimotor approach (i.e. sensorimotor contingencies, virtual presence, etc.) from the experiential level to the material level. It is basically the other side of the coin of the theory of sensorimotor contingencies and the other features of sensorimotor enactivism that we have learned above—it just stresses material realisation rather than experiential realisation. As we have now seen, the sensorimotor approach tracks the *content* and *character* of experience. The dynamical entanglement thesis is an exception to this, since it directly addresses the intertwined material underpinnings that criss-cross the brain–body–environment axis.

The second part of DEUTS, the *unique temporal signature* thesis,[24] adds the *temporal extension* thesis (which was already discussed in the previous section) to the dynamic criss-cross of brain–body–world (the first part of this argument). The interactions mentioned above must evolve over time in order to support experiential states. Without the right kind of temporal evolution—which requires extra-neural elements—the experience cannot be the same. If skilful probing and actions are the basis of perceptual experiences, the basis must be temporal. Probing, acting, touching, and moving are temporal phenomena. Noë captures this idea neatly:

> If this is right, then a neural duplicate of me now, at a moment in time, won't, by dint of being my duplicate now, have any experience at all. If the duplicate does have experience, it will be thanks to its dynamic, temporally extended interaction with the environment. But then again we must note that there is little reason to think that its experience would or could be like mine unless its environment were also like mine. (Noë 2006a, quoted in Clark 2009a, 978)

Thus, taking snapshots at given moments of neural activity won't capture perceptual experience. The whole dynamic, temporally evolving process is needed. And when we add this to the earlier part of this argument (the dynamic entanglement, which implies that we cannot carve off the non-neural from the neural), we get ECM.[25] Perceptual experiences are based on such feedback loops and cannot be reduced to neural factors considered apart from such feedback loops.

Clark himself (2009a, 975–80) thinks that DEUTS is the most promising of the arguments for ECM. This is not surprising, since the dynamical entanglement thesis directly assesses the material underpinnings of perceptual experiences, and that is what ECM is about. In a sense, I agree with him: it is indeed crucial that we reach the sub-personal level. On the other hand, as Ward (2012, 741) notes, the direct link to the sub-personal level implied in the DEUTS argument follows from sensorimotor enactivism's personal-level conception of experience. It is not the primary target for the sensorimotor enactivists themselves when they explain perceptual

[24] "Signature" here refers to the unique temporal evolution (how it has evolved in time) that each experience has.

[25] The phenomenon of temporal extension and the criticism of the snapshot/cinematic view will be discussed in more detail in the next chapter.

experience—but it is a further step. It is the core of ECM and at the same time something that *follows* from sensorimotor enactivism.

Sensorimotor enactivism consists of various theses that I have described above. These aspects are mutually informative and probably cannot be separated. However, if the aim is to name the features that specifically lead to ECM, I think the most important are the following. 1) The theory of sensorimotor contingencies: this ties our phenomenal experiences to our bodily interactions with the environment. Perceptual experiences are constituted by law-like interactions, and this makes it impossible to have a sufficient neural explanation. 2) Bridging the comparative gap: the differences and similarities in the qualities of experiences (within one and across different sensory modalities) are explained by sensorimotor laws—what we *do* with our body, in our environment—therefore, the constitutive basis for the quality of experience comprises external elements. 3) Temporality and situatedness: the dynamic nature of experiences already includes external substrates. 4) Vehicle is determined by content: the external explanation at the personal level requires an external explanation at the sub-personal level (this feature will be discussed in the next section). As we can see, these features show quite unquestionably that if we accept sensorimotor enactivism, ECM follows.

2.3.6 Does Entailment Hold? From the Personal to the Sub-Personal Level

Let us turn our attention to the bridge again. Does it hold up? Does ECM follow from sensorimotor enactivism? Perhaps a possible critique of the bridge could be to claim that sensorimotor contingencies cannot be (part of) the constituents, since they are only (immaterial) *relations*, and as such cannot form the *material* basis for anything. This would, however, be a problematic and mistaken way to understand the conceptions of 'relation' and 'sensorimotor contingency'. Understanding 'relation' without the factors it relates to would be a problematic stance to anybody, including internalists. Wheeler (2015b, 171) enlightens this by remarking that towing is a relative phenomenon between two cars and a rope. Still, it is unquestionable that it occurs in space because the realising vehicles [sic] are located in space. This example shows that even a constitutively relational phenomenon can have a material machinery.

Even though a relational phenomenon, the material basis of sensorimotor contingencies is located in space. A sensorimotor contingency is a mutual interactive relation between the subject and the world by means of sensorimotor action and feedback loops. Those components are intrinsically part of it. The contingency consists of the subject, the extra-neural environmental parts, and the loops between them. Hence, this criticism doesn't work.

Another possible critique would be to argue that sensorimotor skills and actions only causally affect the perceptual experience but do not constitute it. However, the sensorimotorists explicitly state that the relation is a constitutive one. I will not discuss this critique here since the next chapter focuses providing arguments for a constitutive relation.

Moreover, can we move from the personal level to the sub-personal level? Even though sensorimotor enactivism mainly approaches perceptual experience on the experiential (character) level, the phenomenal character gained through the mastery of sensorimotor contingencies has consequences on the material level as well. According to Noë (2004, 221), sensorimotor enactivism entails that "the physical substrate of the experience *may* cross boundaries, implicating neural, bodily and environmental features". Pepper (2014, 99) proposes this claim in an even more precise manner: "sensorimotor dynamics—are physical constituents of perceptual experience". If they are right, it is quite clear that the bridge upholds—if sensorimotor enactivism is true, ECM is also true.

However, Clark (2009a) claims that sensorimotor enactivism is unable to create a firm ground for ECM. He doesn't criticise all aspects of sensorimotor enactivism as such, but instead he denies the implication (or the "second step") of my argument. Is Clark right, and my argument alongside the enactivist argument mistaken?

Ward (2012) has made an important observation: Clark is using a different notion of experience than the enactivists, or rather, they are tracking the phenomenon on different levels. Clark is talking about the material basis of experience and assumes that enactivism is doing the same. However, sensorimotor enactivism approaches experience as a personal-level phenomenon. It is primarily a theory about the qualities of experience (even though the material side is also present, as we will soon see). For example, as mentioned above, Noë's (2004, 2006a) account of virtual presence operates with a personal-level phenomenon—how objects appear to us—it is not even supposed to explain the sub-personal side. Even when sensorimotor enactivists explicitly argue for extension on the sub-personal level (e.g. the argument from dynamic entanglement), it is only due to an inference from their personal-level conception. The sub-personal extension is not sensorimotor enactivism's quintessential argument, but "it is rather a statement of ECM, motivated by their personal level conception of experience" (Ward 2012, 741).

To put it in other words, Clark (mistakenly) attributes a sub-personal approach to sensorimotor enactivism, and yet Clark (correctly) attributes a sub-personal approach to ECM. Hence, ECM and enactivism are targeting the same phenomenon (perceptual experience), but they are targeting it from different levels of explanation. Clark fails to see this difference and instead assumes both of them are operating from the sub-personal level. Nevertheless, he is of course right about the sub-personal-level approach of ECM (as I have stressed from the beginning: ECM is a theory about the material basis of experience).

Sensorimotor enactivism answers *what* is experienced, and ECM answers *how* it is experienced. This use of *whats* and *hows* is borrowed from Hurley's (2010) terminology (and it is nowadays widely used, see, e.g. Wheeler 2015a, b; Kirchhoff and Kiverstein 2019). *What*-explanations tell what the mental state is about, and *how*-explanations tell how they are materially produced. ECM and EM (are answers) to the *how*-question; sensorimotor enactivism is an answer to the *what*-question.

These considerations lead to very profound questions about the nature of the mind itself and how we can investigate it. Even if we corrected the misunderstanding between Clark and the enactivists, the fundamental difference remains: enactivism targets experience as a personal-level phenomenon, whereas ECM is a theory at the sub-personal level, that is, a theory about its material basis. Can the different approaches be combined, or are they just independent theories without a connecting interface?

Sensorimotor enactivists themselves take it for granted that the personal level has implications for the sub-personal level. The *what* determines the *how*. Any other view would be counterintuitive and in conflict with the broader conception of experience that sensorimotor enactivism is committed to. However, as mentioned earlier, in the enactivist literature, this line of reasoning is rather taken for granted than argued for. Ward (2012) makes an exception. He opens up the relation between the personal and sub-personal levels and explains why sensorimotor enactivism's commitment to the implication from the personal to the sub-personal level is warranted. In a word, internalism at the sub-personal level cannot be combined with externalism at the personal level.

> But understanding enactivism instead as involving a relational conception of experience at the personal level, ***with apparent implications for the location of the subpersonal mechanisms of experience***, allows us to make better sense of the enactivist arguments, and make the case for conscious externalism. (Ward 2012, 731, italics added)

Two main reasons for accepting these "apparent implications" between the levels are the following (Ward 2012, 744). On the one hand, there is no intelligible reason why the personal-level explanation that consists of the dynamic interaction between the subject and her environment should shrink to consist only of neural events on the sub-personal level. Experience, straightforwardly, *is* that dynamic relationship, and "the temptation to go internalist when describing experience at the subpersonal level simply will not arise" (Ward 2012, 744). On the other hand, the same applies when we proceed from the other direction: if we were internalists on the sub-personal level, how could experience at the personal level be the dynamic relationship that partly consists of external elements? If the internalist explanation at the sub-personal level were true, the externalist personal-level explanation would be mistaken. Hence, it is easy to see why the enactivists accept ECM as a natural outcome of their view.

> If we adopt the enactivist's conception of experience as a relationship between perceiver and environment then the subpersonal underpinnings of experience must include more than the internal properties of the subject, since those properties leave it underdetermined whether the requisite relationship obtains. (Ward 2012, 746)

To what extent does Clark's criticism of the implication from sensorimotor enactivism to ECM (i.e. the second step of my argument) stand up after these clarificatory remarks on levels? I agree with Ward (2012) that Clark (2009a) is right to refute the implication when both theories are interpreted concerning the sub-personal level. However, when they are understood as concerning the personal level as the sensorimotor enactivists intended (whether or not they clearly expressed it), Clark's critique loses its edge. Hence, sensorimotor enactivism doesn't straightforwardly give us an account of an extended material basis. Rather, it *entails* it; sensorimotor enactivism entails ECM.

Moreover, there is yet another worry presented by Michael Wheeler (2015a, b). He has read his Ward and is well aware of the different levels of explanations in ECM and sensorimotor enactivism (or as he distils the latter, the "relational character of experience"), hence his criticism is potentially even more serious than Clark's. He takes under consideration Ward's claim that the relational character of experience[26] at the personal level determines the sub-personal machinery of those experiences and claims that it faces a dilemma.

According to Wheeler (2015b, 172), one can interpret Ward's view in two ways, but both end up in trouble (from the point of view of a proponent of ECM). First, one can interpret the thesis of the determining relationship as requiring that experience can only happen when there is actual environmental interaction going on. If experiences would always necessitate actual probing, turning one's head and so on, the sub-personal externalism (i.e. ECM) would undoubtedly follow from the relational character of experience. However, as Wheeler observes, this option has the serious side effect that there cannot be perceptual experiences without behaviour. Thus, the first option would lead to an untenable consequence.

According to Wheeler's second possible interpretation, for the determining relationship to obtain, actual environmental probing is not required. It is enough that the perceiver has the access or possibility for environmental interaction: implicit understanding of the outcomes were one to move around and get her hands on the surrounding environment. This time, according to Wheeler (2015b, 173), the problem is that this kind of "knowledgeable readiness to act" is compatible with internalism about the sub-personal machinery of experiences. Thus, the second option would lead to internalism.

However, perhaps the situation for the determining relationship is not as bad as Wheeler depicts. This ostensible dilemma is familiar to us from an earlier discussion in this chapter, namely the distinction between strong enactivism and weak enactivism (although Wheeler doesn't use this terminology or refer to that discussion). Even though these two ways to understand the role of sensorimotor contingencies seem to be in conflict, it doesn't need to be the case. As I said earlier, we cannot cut the sensorimotor skills from the actions themselves, at least when we are talking about humans (and perhaps other animals). The understanding part (weak enactivism) is already embedded in the hands-on, action part (strong enactivism). However, it is not the other way around: online environmental

[26] It simply means that experience is a relationship between experiencer/perceiver and her environment driven by sensorimotor contingencies.

interaction is not necessary for weak enactivism (I will suggest in Chap. 5 that this is one solution to the "dream challenge" for ECM).

It is important to distinguish an online access relation from direct action (or behaviour, to use Wheeler's terminology). Having a (visual) perceptual experience of say, a landscape in front of oneself, doesn't require action as such because one has learned how potential actions would affect it. I suggest that this is how we should interpret strong enactivism, and not as it is interpreted in Wheeler's first option or in Noë's (2016) description of *sensorimotor reductionism* (see discussion in 2.3.3.).

To conclude, my answer to Wheeler is that committing to strong enactivism (in the case of waking-life perceptual experiences) doesn't lead to the "unpalatable consequence that there is no perceptual consciousness without behaviour" (2015b, 172). However, there needs to be an "online access", which means that the agent needs to be attuned to the sensorimotor contingencies that the environment offers. And this is not something that is compatible with internalism. We can embrace strong enactivism (in the case of wakeful perceptual experiences; dreaming and imagining are different matter) without the need to commit to a view that perceptual experience can only occur as a result of concrete behaviour. Therefore, Ward's claim that the relational character of experience at the personal level determines the sub-personal level is not refuted.

2.3.7 *Summing Up the Second Argument*

Of the three arguments presented in this chapter, this one contains the most indisputable bridge from the premise theory to ECM. In the first argument, there was still a way to accept one version of EM (i.e. dispositional EM) without accepting ECM. (Another question is whether anyone really supports dispositional EM alone, but at least that is a way to avoid ECM, if that is the wish.) Also, the implication relation in the third argument has restrictions: all embodiment doesn't always lead to ECM. I claim that the argument at hand differs from the other two: we cannot accept sensorimotor enactivism without accepting ECM because its very core theses inevitably entail ECM.

Besides overlapping, ECM and sensorimotor enactivism are still separate theories because they emphasise slightly different aspects. One central disparity is that sensorimotor enactivism concerns at least *all perceptual experiences*, or also *all sensational experiences* (also, e.g. the sensation of pain), or even *all experience* simpliciter (also non-veridical experiences), depending on the writer. However, my position concerning ECM is a much narrower account that comprises only certain perceptual experiences.

2.4 Third Argument for ECM: The Leaky Body

The third and last argument is based on the embodiment thesis. By embodied experience, I simply mean that the brain is not a minimal sufficient base to create experiences,[27] but the body is part of the constitutive base as well. This much is accepted relatively widely amongst 4E-minded philosophers, enactivists, as well as many phenomenologists. The embodied mind approach (see, e.g. Varela et al. 1991; Hurley 1998; Hurley and Noë 2003; Gallagher 2005, 2018a; Colombetti and Thompson 2008; Cosmelli and Thompson 2010; Thompson and Cosmelli 2011) has been developed in philosophy of mind as well as, for example, in psychology, neuroscience, and developmental robotics. The general claim about the role of the body in explaining cognitive functions is more or less accepted across these fields.[28]

The more debatable part of the argument is the latter part, which is about where does the body end (and the rest of the world begin). According to this second step, our bodies are not always sealed inside our skin, and this opens up a possibility (the third step of the argument) that those external, incorporated parts are partly responsible for the material realisation of some experiences. Thus, the argument is the following. 1) Experience is embodied. 2) The body is sometimes extended to the environment. 3) The incorporated parts sometimes function as substrates for certain experiences.

This argument gives theoretical support for the *functional incorporation* thesis presented in the fourth chapter. The fourth chapter provides

[27] Here, as throughout almost all of this book, by experiences I refer to experiential states rather than the general capability of being (phenomenally) conscious. The distinction has been called, for example state consciousness vs. creature consciousness (see Bayne 2007; Thompson and Cosmelli 2011).

[28] Of course, the view that the mind is realised solely in the brain still has its supporters in the philosophical community (especially amongst analytic philosophers). According to the brain-centred view, the body is not amongst the material realisers for conscious experiences, and it affects our experiences only secondarily by affecting our brain, and in principle an envatted brain could have similar experiences to those of an embodied subject (see criticism of the brain-in-a-vat thought experiment in Chap. 5).

concrete examples of embodiment, incorporation, and crossing the borders of the body, whereas here I will stay at the theoretical level. This argument will concentrate on explaining the argumentative path: why we are allowed to say that experience is embodied, why the borders of the body are extended to the environment, and how all this backs up ECM.

2.4.1 Experience Is Embodied

As a "premise theory" for my argument, the embodiment thesis doesn't require very extensive treatment, since there is already a vast empirical literature about the embodied nature of experiences. The claim that experience is embodied means that (also) the non-neural body is a constitutive basis for perceptual experience. The neural and non-neural parts of the body cannot be separated because they are so intertwined and entangled, and we cannot explain brain functions without acknowledging bodily functions. We can proceed from a negative or positive angle: on the one hand, by criticising the brain-centred, body-deprived view of experience, and on the other hand, by stressing how the body plays a vital role in having experiences. Of course, both ways aim for the same result. In the following, I will use both approaches.

A significant amount of neuroscientific research supports the embodiment thesis (see, e.g. Ferri et al. 2012; Hari and Kujala 2009; Gibbs 2006; Chiel and Beer 1997; Gallese et al. 1996). For example, neuroscientist Riitta Hari considers the brain–body–environment triangle a necessary basis for cognition. According to her, the features brought out in the 4E-framework are a fruitful way to explain the human mind (e.g. Hari 2017). It is widely agreed among neuroscientists that the brain itself is *explanatorily inseparable* from the nervous system and peripheral sensorimotor systems: we cannot understand neural activity without looking at the rest of the body (see Cosmelli and Thompson 2010; Thompson and Cosmelli 2011). As Gallagher (2008, 164) writes, "the structural and functional design of the body shapes the way that we experience the world".

The nervous system constantly receives and responds to the feedback from the body, its movement, and its environment, and action–perception loops are essential for a simple movement as well as for more complex cognitive functions (Hari 2017; Chiel and Beer 1997). Cosmelli and Thompson (2010, 12) give a list of functional systems that are dependent on brain–body coupling. They include the interoceptive system, vestibular autonomic regulation, balance and somatic graviception, and interaction between the senses occurring at central and peripheral levels.

An example that is directly connected to perceptual experience is empirical research on vision. Studies in active or animate vision (Ballard 1991; Ballard et al. 1992) has revealed that eyes don't passively record what is in the scene, but eye saccades and head motions (which are, obviously, bodily actions) are shown to take part in creating the visual experience. Maintaining visual perception is dependent on microsaccades (short, involuntary eye movements): if they are prevented, object perception will fail (Martinez-Conde et al. 2000). Overall, neuroscientific research on vision suggests that visual experience is a strongly embodied phenomenon, dependent on the subject's voluntary and involuntary bodily actions and on connected environmental loops.[29]

Recently, there has been a lot of discussion about the embodied nature of affective states, such as emotions and moods (see, e.g. Colombetti and Thompson 2008). The philosophical arguments for embodied affectivity are backed up by a vast amount of empirical research (see overviews in Gibbs 2006; Colombetti and Torrance 2009). For example, facial expressions, gestures, and autonomic nervous system activity are partly responsible for affective states. Dimberg et al. (2000) discovered that the non-conscious perception of facial expressions (carried out with the backward masking technique) affect the perceiver's facial expressions. For example, pictures of sad expressions increased activation in the muscle that is usually connected with negative emotions (the *corrugator supercilii* muscle), even though input sad expressions were hidden with backward masking and hence not consciously experienced.

The embodiment of emotions is starting to be quite widely accepted among both philosophers of emotion and empirical scientists. Although more controversial, the claim that affective states extend *outside* the skin is also being supported by a growing number of researchers (e.g. Colombetti 2014; Colombetti and Roberts 2014; Krueger 2018). Although this line of research does not target exactly the same phenomenon as I am here (i.e.

[29] Another example of the importance of the body for cognitive and affective states and information processing located outside the brain is the enteric nervous system, "brain in the gut". However, it is not straightforwardly connected to perception, as opposed to, for example, information processing in the retina.

perceptual experience), it is clear that the step from (phenomenal) affective experiences to (phenomenal) perceptual experiences is a short one.[30]

One example of embodiment is the role of bodily gestures and expressions. Bodily gestures are not neural activity, but they are still part of cognitive processing. Goldin-Meadow (2003), based on her thorough experiments about human gestures, shows that gestures do not only express internal states but have a functional role in thinking; they are part of thinking itself. In an experiment, two groups of children were given a list to memorise, and then they needed to solve a mathematical problem before trying to remember the list again. One group could gesture as they wished during the maths task, but the other group was prevented from gesturing. The results were clear: the freely gesturing group managed significantly better in the task of remembering the list (Goldin-Meadow et al. 2001; discussed e.g. in Clark 2008). Based on this and other experiments, Goldin-Meadow (2001, 2003) ends up concluding that gesturing is not only a motor act separate from cognitive functions but part of the cognitive process itself.

These examples only scratch the surface of the empirical and philosophical work that has been conducted in recent years that shows that our cognitive and experiential faculties are deeply embodied. Consequently, we could say that the embodiment of experiences is no longer a highly controversial claim but rather something that a significant part of the research community is starting to accept.

2.4.2 *The Frontiers of the Body Are Not Rigid*

The second step of this argument is that the body is "leaky": it is not always enclosed inside the skin, but it can contain environmental elements. I will distinguish the fine-grained details of different versions of bodily extensions and incorporation in the fourth chapter (from mere extension

[30] One more way to argue for ECM would be to start from extended affectivity and then infer that because affective states are phenomenal experiences, if they are extended, phenomenal consciousness is already extended. As things stand, the hardest part to accept in ECM is undoubtedly the conscious, phenomenal side, and not the fact that it is about *perceptual* experiences. I haven't seen this kind of argument for extended experience in the literature, though. The proponents of extended affectivity (most importantly Colombetti, Roberts, and Krueger) are not trying to argue for ECM as such. It is not their target, but I think ECM follows from their account, nevertheless. Their examples (e.g. a violinist and her violin) are similar to incorporating non-biological tools into the biological body.

to functional incorporation), whereas here I discuss in a more general way the claim that the body takes external objects into it.

When we use a stick to reach an apple from a tree, that stick has become an extension of our hand—we stop feeling the stick in our hand, and instead we feel the apple through our stick. There is neuroscientific evidence that a stick can become part of the functional body in situations where we reach for something (see, e.g. Maravita and Iriki 2004; Berti and Frassinetti 2000). Depending on how we see it, we can either say that the body leaks into the environment or say that it incorporates environmental objects—but the outcome is quite the same.

The notion of 'body' is somewhat ambiguous. Its frontiers may prove to be different depending on what we are looking for. For example, we can refer to *body schema* (body as an object) or to *body image* (body as a subject). (This distinction is explored further in Chap. 4.) A central question for this argument is, of course, 'is biology the crucial factor?' What about an artificial knee, a dental filling, or a pacemaker? What about biological non-human elements located in the body, such as bacteria? What about prostheses, clothes, piercings, hair, wigs, nails, artificial nails, eyeglasses, contact lenses, ocular prostheses, and others? Or microplastic that humans (and animals) swallow a considerable amount of every day?

It would seem that biology cannot be the defining criterion since our bodies contain several crucial non-biological components that replace or enhance biological functions. Our bodies are hybrid systems. We are "natural-born cyborgs" or "human-technology symbionts" as Clark (2003) describes. The human body is not a fixed system, but flexible, constantly changing, and incorporating external resources. It is made of both biology and technology, and the borders of the body are negotiable.

If one were reliant on, for example, eyeglasses, a walking stick, a blind person's white cane, a hearing aid, or sensory substitution devices, their loss would be somewhat comparable to losing one's body part. Of course, the artificial devices would be easier to substitute with new ones, but in principle biological and non-biological parts play the same role: they are contingent, replaceable. Not even our biological body stays the same: it is different from what it was ten years ago. All the cells (brain cells and other kinds of cells) have been replaced with newly regenerated cells. Instead of a fixed substrate, it is the functions and interactions between the brain, the body, and the world that create experience.

2.4.3 *External Body Parts as Substrates of Experience*

If we agree with the first and second steps of this argument, we still need to ask, 'are those external elements part of the constitutive base for experiences?' After all, it seems possible to accept that experience is embodied and that the body is not always defined by the skin and yet deny that the external parts could be part of the constitutive base for experiences. Indeed, the second step doesn't automatically lead to the third step because it is possible to hold that bodies may have external parts but that those parts are only externalisations of the body, but not material realisers of experiential states. However, this line of thought would undermine the first and second steps of the argument. For, if we commit to embodied realisation of certain experiences, and to our capability of incorporating non-biological props into our bodily sensorimotor functions (i.e. into the embodied substrate of those experiences), the third step becomes difficult to deny.

As several empirical experiments have shown (e.g. Maravita and Iriki (2004) about training macaque monkeys to use rakes; Bach-y-Rita and Kercel (2003) about sensory substitution devices), when used in the right way, the external parts will gain a new functional and phenomenological status. For example, when a blind person uses a cane, it is no longer an external object for her but gradually becomes part of her point of view (see Chap. 4 for more elaborated examples). They are no longer merely external objects but become part of the sensing body—part of the (quasi-transparent) subjective point of view rather than an external object. This new status is the reason why they cannot simply be overlooked when searching for the realisers of experience. As we have learned in the first two arguments for ECM, experience *is* that embodied, sensorimotor action in the environment, and that action is partly realised by those incorporated objects. Incorporated tools have their seat at the interface of the body and environment, and if we accept that both the body and the environmental action are parts of the substrate of experiences, there are no good reasons why we should stop short and forbid the mediating element (namely the incorporated object) this role.

Of course, not all kinds of bodily extension will suffice. The object needs to be a certain kind of tool with a sensorimotor function, and it needs to gain a new phenomenological status in use (e.g. a blind cane), differentiated from a mere tool (e.g. a washing machine) or a mere extension of a material body (e.g. an artificial knee). (For a more detailed description of the criteria and a taxonomy of different levels of externalisation and incorporation, see Chap. 4.)

The third argument was yet another way to reach the conclusion that external parts can be part of the material realisers for experiences. When we allow that our bodily functions can be partly responsible for some experiential states, it is a small step to the environment. If an action carried out with a hand is included, it is a slippery slope to the tool that the hand is holding. With our bodies extending to the external parts, our experiential states are also sometimes materially realised by those externalisations. A leaky body means a leaky mind.

2.5 Conclusions to the Three Arguments

These arguments have shown that in order to stay consistent, a supporter of the 4E-framework should accept ECM as well. As the reader might have noticed, the three arguments are not independent of each other, but they are different aspects in a common ground. They can be taken as a cumulative argument for ECM, where the three arguments build on each other: the argument from sensorimotor enactivism and the argument from occurrent EM lead to the argument from embodiment. Each of them, however, also offers a distinct entry to ECM.

The first argument, based on inference from EM to ECM, gives the groundwork for the concept of experience that will be used (and still developed) especially in the second argument. The second argument, based on sensorimotor enactivism, is in a sense the most important argument of them all, because the premise theory itself targets extended experience in a straightforward manner. The third argument about the embodied experience and the leaky body (and hence leaky experience) ends up with a conclusion that is not new but has nevertheless been stressed in a wide range of philosophical and empirical research about sensing with tools (see, e.g. De Preester and Tsakiris 2009; Maravita and Iriki 2004). In this argument, my way of framing the matter is, however, different from how it is usually presented and is also different from how I present the same matter in other chapters (especially in Chap. 4). I haven't seen this kind of approach in arguing for ECM in the 4E literature.

No doubt these arguments for ECM might be hard to accept for analytic philosophers of mind, because mental phenomena have traditionally been described in such a different way. There is resistance among philosophers of mind as a consequence of the strict separation between cognition (understood as computation) and consciousness (understood as something that doesn't fit the computationalist picture). Empirical scientists don't have the conceptual burden of bats, zombies, Chinese rooms, explanatory gaps, and hard problems of consciousness, and thus these arguments might be easier to accept for them.

References

Aizawa, K. (2010a): "Consciousness: don't give up on the brain". *Royal Institute of Philosophy Supplement*, 67, 263–284.

Bach-y-Rita, P., & Kercel, S.W. (2003): "Sensory Substitution and the Human-Machine Interface". *Trends in Cognitive Sciences* 7: 12, 541–546.

Ballard, D. H. (1991): "Animate Vision". *Artificial Intelligence*, 48(1), 57–86.

Ballard D. H., & Brown C., M. (1992): "Principles of animate vision". *CVGIP: Image Understanding*, 56(1): 3–21.

Bayne, T. (2007): "Conscious States and Conscious Creatures: Explanation in the Scientific Study of Consciousness". *Philosophical Perspectives* 21, 1–22.

Berti, A., & Frassinetti, F. (2000): "When far becomes near: Re-mapping of space by tool use". *Journal of Cognitive Neuroscience* 12, 415–420.

Block, N. (1978): "Troubles with functionalism". *Minnesota Studies in the Philosophy of Science* 9, 261–325.

Block, N. (2002): "Concepts of Consciousness". In Chalmers (ed.): *Philosophy of Mind, Classical and Contemporary Readings*. New York: Oxford University Press, 2002, 206–218.

Bruno, M. G. (2013): *The Extended Conscious Mind*. University of Arizona. (PhD-dissertation).

Brooks, R. A. (1990): "Elephants don't play chess". *Robotics and Autonomous Systems*. 6(1–2), 139–159.

Chalmers, D. (1996): *The Conscious Mind: In Search of a Fundamental Theory*. New York: Oxford University Press.

Chalmers, D. (2002): "Consciousness and Its Place in Nature". In Chalmers, D. (ed.): *Philosophy of Mind, Classical and Contemporary Readings*, New York: Oxford University Press, 2002, 247–272.

Chalmers, D. (2008): "Foreword". In Clark, A.: *Supersizing the Mind: Embodiment, Action and Cognitive Extension*. Oxford, UK: Oxford University Press.

Chemero A. (2009): *Radical Embodied Cognitive Science*. Cambridge, MA: MIT Press.

Chemero, A. (2016): "Sensorimotor Empathy". *Journal of Consciousness Studies*, 23, 138–152.

Chiel, H. & Beer, R. (1997): "The brain has a body: Adaptive behavior emerges from interactions of nervous system, body and environment". *Trends in Neurosciences*, 20, 553–57.

Clark, A. (2003): *Natural-Born Cyborgs: Minds, Technologies, and the Future of Human Intelligence*. Oxford, UK: Oxford University Press.
Clark, A. (2007a): "Curing cognitive hiccups: a defence of the extended mind". *Journal of Philosophy*, 104, 163–192.
Clark, A. (2007b): "Re-Inventing Ourselves: The Plasticity of Embodiment, Sensing and Mind". *Journal of Medicine and Philosophy*, 32: 263–282.
Clark, A. (2008): *Supersizing the Mind: Embodiment, Action and Cognitive Extension*. Oxford, UK: Oxford University Press.
Clark, A. (2009a): "Spreading the joy? Why the machinery of consciousness is (probably) still in the head". *Mind*, 118(472), 963–993.
Clark, A. (2010a): "Memento's Revenge: The Extended Mind, Extended". In Menary, R. (ed.), *The Extended Mind*. Cambridge, MA: MIT Press, 43–66.
Clark, A. (2012): "Dreaming the whole cat: Generative models, predictive processing, and the enactivist conception of perceptual experience". *Mind*, 121(483), 753–771.
Clark, A. (2013): "Whatever next? Predictive brains, situated agents, and the future of cognitive science." *Behavioral and Brain Sciences*, 36 (3), 181–204.
Clark, A. & Chalmers, D. (1998): "The Extended Mind". *Analysis* 58: 1, 7–19.
Colombetti, G. (2014): *The Feeling Body: Affective Science Meets the Enactive Mind*. Cambridge, Massachusetts: The MIT Press.
Colombetti, G. & Thompson, E. (2008): "The Feeling Body: Toward an Enactive Approach to Emotion". In Overton W., Muller U., & Newman J. (eds.): *Developmental Perspectives on Embodiment and Consciousness*. New York: Lawrence Erlbaum, 45–68.
Colombetti, G. & Torrance, S. (2009): "Emotion and ethics: An inter-(en)active approach". *Phenomenology and the Cognitive Sciences* 8, 505–526.
Colombetti, G. & Roberts, T. (2014): "Extending the Extended Mind: The Case for Extended Affectivity". *Philosophical Studies* 172 (5): 1243–1263.
Colombetti, G. & Krueger, J. (2015): "Scaffoldings of the affective mind". *Philosophical Psychology*. https://doi.org/10.1080/09515089.2014.976334
Cosmelli, D. & Thompson, E. (2010): "Embodiment or Envatment? Reflections on the bodily basis of consciousness". In J. Stewart, O. Gapenne & E. Di Paolo (eds.), *Enaction: Towards a New Paradigm for Cognitive Science*. MIT Press.
Degenaar, J., & O'Regan, K. (2015): "Sensorimotor Theory and Enactivism". *Topoi*, 36:3, 393–407. https://doi.org/10.1007/s11245-015-9338-z
De Preester, H. & Tsakiris, M (2009): "Body-extension versus body-incorporation: Is there a need for a body-model?" *Phenomenology and the Cognitive Sciences*, 8(3): 307–319.
Dewey, J. & Bentley, A. (1949): *Knowing and the Known*. Boston: Beacon Press.
Dimberg, U., Thunberg, M., & Elmehed, K. (2000): "Unconscious facial reactions to emotional facial expressions". *Psychological Science*, 11, 86–89.
Di Paolo, E. (2009): "Extended Life". *Topoi* 28: (1), 9–21.

Ferri F., Frassinetti F., Ardizzi M., Costantini M. & Gallese V. (2012): "A sensorimotor network for the bodily self". *Journal of Cognitive Neuroscience*, 24: 1584–95. https://doi.org/10.1162/jocn_a_00230

Gallagher, S. (2005): *How the Body Shapes the Mind*. New York: Oxford University Press.

Gallagher, S. (2008): "Intersubjectivity in perception". *Continental Philosophy Review*, 41: 163–178.

Gallagher, S. (2018a): "New Mechanisms and the Enactivist Concept of Constitution". In M. P. Guta (ed.): *The Metaphysics of Consciousness*, 207–220. London: Routledge.

Gallese, V., Fadiga, L., Fogassi, L. & Rizzolatti G. (1996): "Action recognition in the premotor cortex," *Brain*, 119: 593–609.

Gangopadhyay, N. & Kiverstein, J. (2009): "Enactivism and the Unity of Perception and Action". Topoi 28 (1): 63–73.

Gibbs, R.W. (2006): *Embodiment and Cognitive Science*. Cambridge: Cambridge University Press.

Gibson, J. J. (1979): *The Ecological Approach to Visual Perception*. Boston, Houghton Mifflin.

Goldin-Meadow, S., Nusbaum, H., Kelly, S. D, & Wagner S. (2001): "Explaining math: Gesturing lightens the load". *Psychological Science* 12(6): 516–22.

Goldin-Meadow, S. (2003): *Hearing Gesture: How Our Hands Help Us Think*. Cambridge, MA: Harvard University Press.

Hari, R. (2017): "From Brain-Environment Connections to Temporal Dynamics and Social Interaction: Principles of Human Brain Function". *Neuron*, 7:94, 1033–1039. https://doi.org/10.1016/j.neuron.2017.04.007.

Hari, R. & Kujala, M. V. (2009): "Brain basis of human social interaction: from concepts to brain imaging". *Physiol. Rev.*, 89, 453–479.

Hurley, S. (1998): *Consciousness in Action*. Cambridge, MA: Harvard University Press.

Hurley, S. (2010): "The Varieties of Externalism". In Menary, R. (ed.), *The Extended Mind*. Cambridge, MA: MIT Press, 101–153.

Hurley, S. & Noë, A. (2003): "Neural Plasticity and Consciousness". *Biology and Philosophy*, 18(1), 131–168.

Hutto, D. & Myin, E. (2012): *Radicalizing Enactivism: Basic Minds without Content*. Cambridge, MA: The MIT Press.

Hutto, D. & Myin, E. (2017): *Evolving Enactivism: Basic Minds Meet Content*. Cambridge, MA: The MIT Press.

Kirchhoff, M. D. & Kiverstein, J. (2019): *Extended Consciousness and Predictive Processing: A Third Wave View*. London, UK: Routledge.

Kiverstein, J. (2010): "Sensorimotor knowledge and the contents of experience". In Gangopadhay, N., Madary M. & Spicer, F. (eds.): *Perception, Action, and Consciousness*. Oxford University Press, 257–274.

Kiverstein, J. (2016): "The Interdependence of Embodied Cognition and Consciousness". *Journal of Consciousness Studies*, 23(5–6): 105–137.

Kiverstein, J. & Clark, A. (2009): "Introduction: Mind Embodied, Embedded, Enacted: One Church or Many?" *Topoi*, 28(1), 1–7.

Krueger, J. (2018): "Schizophrenia and the Scaffolded Self". *Topoi*. https://doi.org/10.1007/s11245-018-9547-3

Levine, J. (1983): "Materialism and qualia: The explanatory gap". *Pacific Philosophical Quarterly*, 64, 354–361.

Loughlin, V. (2012): "Sketch this: Extended mind and consciousness extension". *Phenomenology and the Cognitive Sciences*, 12, 41–50.

Maiese, M. (2015): *Embodied Selves and Divided Minds*. New York, Oxford University Press.

Maravita, A., & Iriki, A. (2004): "Tools for the body (schema)". *Trends in Cognitive Sciences*, 8(2), 79–86.

Martinez-Conde, S., Macknik, S. L. & Hubel D. H. (2000): "Microsaccadic eye movements and firing of single cells in the striate cortex of macaque monkeys". *Nature Neuroscience*, 3(3): 251–258.

Menary, R. (2006): "Attacking the bounds of cognition". *Philosophical Psychology*, 19, 329–344.

Menary, R. (2007): *Cognitive Integration: Mind and Cognition Unbounded*. New York: Palgrave Macmillan.

Menary, R. (ed.) (2010a): *The Extended Mind*. Cambridge, MA: MIT Press.

Menary, R. (2010b): "Introduction: The Extended Mind in Focus". In Menary R (ed.) *The Extended Mind*. Cambridge, MA: MIT Press, 1–26.

Menary, R. (2010c): "Cognitive Integration and the Extended Mind". In Menary R (ed.) *The Extended Mind*. Cambridge, MA: MIT Press, 227–244.

Noë, A. (2004): *Action in Perception*. Cambridge, MA: MIT Press.

Noë, A. (2006a): "Experience without the head". In Gendler, T. S. and Hawthorne, J. (eds.) *Perceptual Experience*. New York: Oxford University Press. 411–434.

Noë, A. (2008): "Magic Realism and the Limits of Intelligibility: What Makes Us Conscious?" In Hawthorne, J. (ed.) *The Philosophy of Mind*. Oxford: Blackwell. 457–476.

Noë, A. (2012): *Varieties of Presence*. Cambridge, MA: Harvard University Press.

Noë, A. (2016): "Sensations and Situations: A Sensorimotor Integrationist Approach". *Journal of Consciousness Studies*, 23, No. 5–6, 66–79.

O'Regan, K. (2011): *Why red doesn't sound like a bell: Understanding the feel of consciousness*. New York: Oxford University Press.

O'Regan, K. (2016): "How the Sensorimotor Approach to Consciousness Bridges Both Comparative and Absolute Explanatory Gaps". *Journal of Consciousness Studies*, 23, 5–6, 39–65.

O'Regan, K., Rensik, Clark, J. (1999): "Change blindness as a result of mud-splashes". *Nature* 398: 34.

O'Regan, K. & Noë, A. (2001): "A sensorimotor account of vision and visual consciousness". *Behavioral and Brain Sciences.* 24(5): 939–973.
O'Shaughnessy, B. (2000): *Consciousness and the World.* New York: Oxford University Press.
Pepper, K. (2014): "Do sensorimotor dynamics extend the conscious mind?" *Adaptive Behavior*, 22(2), 99–108.
Prinz, J. (2015): "Unconscious Perception". In Matthen, M. (ed.) *The Oxford Handbook of Philosophy of Perception*, Oxford: Oxford University Press, 371–392.
Pritchard, D. (2010): "Cognitive ability and the extended cognition thesis". *Synthese*, 175, 133–151.
Putnam, H. (1975): "The Meaning of 'meaning'". In Gunderson, K. (ed.), *Language, Mind, and Knowledge. Minneapolis*: University of Minnesota Press.
Rensink, R. A., O'Regan, J. K., & Clark, J. J. (1997): "To see or not to see: The need for attention to perceive changes in scenes". *Psychological Science*, 8, 368–373.
Searle, J., R. (1992): *The Rediscovery of Mind.* Cambridge, MA: MIT Press/ Bradford Books.
Seth, A. K., Verschure, P. et al. (2016): "Action-Oriented Understanding of Consciousness and the Structure of Experience". In Engel, A., Friston, K., & Gragic, D. (eds.) *The Pragmatic Turn: Toward Action-Oriented Views in Cognitive Science.* Strüngmann Forum Reports 18, The MIT Press, 261–284.
Silberstein M. & Chemero A. (2011): "Complexity and Extended Phenomenological-Cognitive Systems". *Topics in Cognitive Science*, 4 (1), 35–50.
Silberstein M. & Chemero A. (2015): "Extending Neutral Monism to the Hard Problem". *Journal of Consciousness Studies*, 22 (3–4), 181–194.
Silverman, D. (2013): "Sensorimotor enactivism and temporal experience". *Adaptive Behavior* 21(3), 151–158.
Simons, D. & Chabris, C., (1999): "Gorillas in our midst: Sustained inattentional blindness for dynamic events". *Perception*, 28(9), 1059–1074.
Stafford, T. & Webb, M. (2005): *Mind Hacks.* Sebastopol, CA: O'Reilly.
Strawson, G. (1994): *Mental Reality.* Cambridge, MA: MIT Press/Bradford Books.
Sur, M., Angelucci A., & Sharma J. (1999): "Rewiring Cortex: The Role of Patterned Activity in Development and Plasticity of Neocortical Circuits". *Journal of Neurobiology*, 41, 33–43.
Sutton, J. (2010): "Exograms and Interdisciplinarity: History, the Extended Mind, and the Civilizing Process". In Menary, R. (ed.) *The Extended Mind.* Cambridge, MA: MIT Press, 2010, 189–226.
Taylor, J. (1962): *The Behavioral Basis of Perception.* New Haven and London: Yale University Press.
Thompson, E. (2007): *Mind in Life: Biology, Phenomenology, and the Sciences of Mind.* Cambridge, MA: Harvard University Press.

Thompson, E. & Cosmelli, D. (2011): "Brain in a Vat of Body in a World? Brainbound versus Enactive Views of Experience". *Philosophical Topics* 39(1), 163–180.

Varela F., Thompson E., & Rosch E. (1991): *The Embodied Mind: Cognitive Science and Human Experience*. Cambridge, MA: The MIT Press.

Vold, K. (2015): "The parity argument for extended consciousness". *Journal of Consciousness Studies* 22: 16–33.

Ward, D. (2012): "Enjoying the spread: Conscious externalism reconsidered". *Mind*, 121(483), 731–751.

Ward, D., Silverman, D., & Villalobos, M. (2017): "Introduction: The Varieties of Enactivism". *Topoi* 36, 365–375.

Wheeler, M. (2010): "In defence of extended functionalism". In Menary, R. (ed.) *The Extended Mind*. Cambridge, MA: MIT Press, 245–270.

Wheeler, M. (2015a): "Not What it's Like but Where it's Like: Phenomenal Consciousness, Sensory Substitution, and the Extended Mind". *Journal of Consciousness Studies*, 22(3–4): 129–147.

Wheeler, M. (2015b): "Extended Consciousness: an Interim Report". *The Southern Journal of Philosophy. Spindel Supplement: Alternative Models of the Mind*. Vol. 53, 155–175.

Wilson, R. A. (2004): *Boundaries of the Mind: The Individual in the Fragile Sciences: Cognition*. New York: Cambridge University Press.

Wilson, R. A. (2010): "Meaning making and the mind of the externalist". In Menary, R. (ed.) *The Extended Mind*. Cambridge, MA: MIT Press, 167–188.

CHAPTER 3

The Constitution-Turn and Extended Conscious Mind

3.1 Introduction

This chapter discusses how to differentiate causal from constitutive relations in general, and specifically in EM and ECM. The focus is on the *concept of constitution*, whereas Chap. 4 describes the constitutive relation specifically in ECM and how to detect it and separate from causal background conditions. The demand for these discussions stems from what I call the *constitution-turn* in the extended mind literature.

Starting from Adams and Aizawa's causal–constitution objection presented shortly after the original paper of Clark and Chalmers (1998), there has been an increased interest towards the concept and ontology of constitution as a culminating question whether externalist views in philosophy of mind are true. Especially in recent years it has gained a great deal of attention in the literature. As Baumgartner and Wilutzky (2017, 1105) put it, to escape the stalemate situation about the adequate *mark of the cognitive*, the EM literature has turned into a debate about an adequate *mark of constitution*. However, an account of constitution in EM cannot be used directly for ECM since they have such differing views on the

P. Telakivi, *Extending the Extended Mind*, New Directions in Philosophy and Cognitive Science,
https://doi.org/10.1007/978-3-031-35624-7_3

phenomenon they are explaining. That is why ECM needs its own treatment, and the main goal of this chapter is to provide that.[1]

By clarifying the concepts of constitution and causality, this chapter will respond not only to the *causal–constitution fallacy* but also to a closely related critique, that is, the *cognitive bloat argument* against EM and ECM. I argue that these criticisms arise from a mistaken concept of constitution that should be used with ECM (and occurrent experiences in general), and they can be resolved by analysing and clarifying the notion of constitution.

3.1.1 The Causal–Constitution Fallacy

The causal–constitution distinction is at the centre of the debate concerning externalism in philosophy of mind. The extension (what is meant by *extended* in the 4E's) in both EM and ECM is a *constitutive relation* according to the advocates of EM and ECM.[2] The interaction loops with external objects and devices partly constitute cognition in EM and experience in ECM. The external parts are constitutive components[3] of cognition/experiences. This is what *extended* refers to in EM and ECM. The external and internal components together constitute the material basis for some mental states, and together they are necessary and sufficient for those states.[4]

However, it has been argued (e.g. Adams and Aizawa 2001, 2010; Aizawa 2010) that EM has confused causal and constitutive relations, that is, committed a *causal–constitution fallacy*, also known as a *coupling–constitution fallacy* (C–C fallacy, from now on).[5] According to this objection,

[1] To my knowledge, Kirchhoff and Kiverstein (2019) were the first ones to deal with the metaphysics of constitution in the context of ECM. They also combine the predictive processing framework to their account.

[2] As was stated in the Chap. 1, this is what differentiates *extended* from *embedded*—the latter stresses the causal dependency on the environment but denies the constitutive extension outside the boundaries of the organism.

[3] By 'components', I refer to 'constitutive components', and I use the terms 'component' and 'constituent' interchangeably throughout this text.

[4] The brain is necessary (at least for present-day human minds), but not always sufficient condition for cognition (according to EM) and for experiences (according to ECM).

[5] I use the term 'coupled' as neutral regarding the causal–constitution question. So if p is coupled with q, we still need to ask whether that relation is a causal or a constitutive one. The use of the term is not consistent in the literature; sometimes it is used synonymously with causal relations, sometimes with constitutive relations.

the role of external elements is at best causal, but certainly not constitutive: "the fact that object or process X is coupled to object or process Y does not entail that X is part of Y" (Adams and Aizawa 2010, 68). From the fact that environmental phenomena causally affect our cognition, the proponents of EM mistakenly (according to the C–C fallacy objection) inferred that the causally affecting relation is (or leads to) a constitutive relation. Thus, the C–C fallacy objection aims to be a reminder that one should not draw constitutive conclusions from causal premises. My aim is to show that neither EM nor ECM has committed this error, and the C–C fallacy objection can be rejected.

Adams and Aizawa's original formulation of the critique (which concerns how Clark and Chalmers use the Otto example) is relatively easily solved because they in the first place have misunderstood Clark and Chalmers's intentions about EM. Adams and Aizawa interpret EM's question to be thus: under what conditions does a physical object become cognitive? However, EM does not claim that coupling would make any object *itself* cognitive. Rather, EM is asking—in Clark's (2010, 84) own words—"when is some physical object or process part of a larger cognitive system?"

In their criticism of EM, Adams and Aizawa appeal to the *mark of the cognitive*, by which they refer to the manipulation of representations that bear intrinsic, non-derived content. Allegedly, external parts do not bear non-derived content, and therefore they don't satisfy the requirement for the mark of the cognitive. "Where the symbols written in Otto's notebook have merely derived content, the recollection in Inga's brain has non-derived content" (Adams and Aizawa 2001, 55). It is, however, very unclear what non-derived content actually is, why cognitive states should have it, or why external parts should *a priori* lack it.[6] Menary (2010b) has noted that the distinction between derived and non-derived content is not relevant in cognitive sciences and that it has no value in scientific research on cognition. (For a thorough discussion about Adams and Aizawa's original critique, see Menary 2010a.)

[6] Even though I think pleading for non-derived content as the mark of the cognitive is not convincing to start with, if for the sake of argument we accepted it, ECM would actually survive the attack better than EM. It is after all evident that perceptual experiences involve less derivation than some standard examples of dispositional EM. (Of course, Adams and Aizawa never intended their critique to be directed towards ECM.)

Nevertheless, there is a serious and unsettled matter underlying the original critique that deserves a proper examination, namely the question whether the relation in EM actually is a causal or constitutive one, and how we distinguish causal from constitutive. The critique was originally directed towards EM, but if it is a sound critique, it also applies to ECM. If correct, the C–C fallacy is as serious a worry for ECM as it is for EM.[7]

Furthermore, Clark (2009) has introduced a more empirically based version of the C–C fallacy objection specifically against ECM (for a more detailed discussion about Clark's critique, see Chap. 5). Adams and Aizawa's critique is drawn from metaphysical considerations (viz. the metaphysical assumption of the mark of the cognitive and the failure of external parts to satisfy it as a reason why EM is not possible), whereas Clark is appealing to the superior processing power of the biological brain as a reason why ECM is not possible.[8] Accordingly, the former is *a metaphysical argument for the C–C fallacy objection*, and the latter is *an empirical argument for the C–C fallacy objection*. The C–C fallacy objection I am dealing with in this chapter is the metaphysical version. It does originate from Adams and Aizawa's writings, but I treat it as a general opposition for the constitutive claim. Even though the whole setting is rather metaphysical, it doesn't rule out empirical investigation. Kirchhoff (2015) writes:

> [B]y scrutinizing the metaphysics of what it means for certain Xs to compose a certain Y now, it is possible, I think, to turn what might look like a metaphysical dispute into a productive recipe for empirical research and to set certain constraints for how such research must be carried out.

The constitution-turn has raised the causal–constitutive distinction into the centre of the debate whether the arguments for different versions of extended mind are sound: the proponents claim that the relation is constitutive, and the opponents claim that it is merely causal. However, problems arise because neither side has clarified what they actually mean by this distinction, and that has created confusion. But the confusion itself is not a reason to attack externalist views, but rather it is an ambiguity that all parties need to address and find a solution to. In order to be able to respond to the C–C fallacy objection (or to keep on using it as a critique

[7] There are, however, some differences in how the two can respond to the criticism. I will get back to this later.

[8] Both Clark and Adams and Aizawa admit that their internalism is a contingent rather than necessary fact: it is internalism about us "contemporary humans".

towards externalist views), we need to get a clearer picture of the distinction between causal and constitutive relations. The clarification might help the discussions about not only EM and ECM but also philosophy of mind in general and perhaps the general philosophical question of the concepts of causality and constitution as well, for example, by broadening the view of what counts as a constitutive relation.

Answering the objection leads us to a very profound question of what belongs to cognition and experience, and how to set their limits. Marie Kaiser (2018, 5–6) cuts up the question and reveals that these kinds of demarcations actually consist of slightly different sides, for example, the *individuation of the whole question* (what is the whole and how it should be identified) is different from the *demarcation question* (where are the boundaries of the whole, and how it can be differentiated from its environment).[9] This can be compared with a distinction between *operational limits* (cf. individuation of the whole) and *spatial boundaries* (cf. demarcation) (Virgo et al. 2011). These two demarcation criteria don't need to coincide: the material boundaries can be different from those of the organismic boundaries. EM literature has focused mainly on the question of material boundaries, even though the question about operational/organisational limits could (and should) be raised as well. In order to set the limits, we need answers to both of these kinds. (This question of course motivates this whole book, not only this chapter.)

We need to find a way of setting the boundaries that is "neither too wide nor too narrow, but rather 'just right'", as Weiskopf (2010) has framed the problem (and accordingly called it the "Goldilocks problem"). Of course, what counts as "just right" is a highly complicated question. As Kaplan (2012, 551) has put it, we need to "contend with the problem of partitioning from among the set of causally relevant factors those that are legitimate components of a given mechanism or system from those that play roles as part of the causal background". This is a question worth asking and trying to answer, and the task is not an easy one. Much depends on how we understand the notion of constitution in the first place, how we distinguish it from causation, and how these concepts are applied to EM/ECM frameworks.

[9] Kaiser (2018) is discussing parthood, demarcation, individuation, and decomposition in biology. Nevertheless, the same differentiation can (and should) be kept in mind when we are dealing with cognition and experience.

3.1.2 *The Constituted Phenomenon: The Dynamic Nature of Experiences*

Before delving into the concept of constitution, let us have a look at the phenomenon in which the constitution relation allegedly occurs (viz. perceptual experience). As was already presented in the second argument for ECM in Chap. 2, the nature of experiences is *dynamic*: they are not static states but rather unfolding over time by means of constant dynamic change.[10] This view of experience is influenced by enactivism, according to which perceptual experience is a temporally extended interaction between the agent and its environment governed by sensorimotor contingencies (O'Regan and Noë 2001; Noë 2004, 2006a; Buhrmann et al. 2013; Silverman 2013; Pepper 2014). Experience is not made up of a succession of separate atom-like instances but instead evolves dynamically between the agent and its surroundings. David Silverman (2013, 154) encapsulates sensorimotor enactivism's idea about temporal experience as follows:

> [T]he physical substrate of the experience is a smoothly continuous activity rather than one which breaks down into temporally discrete chunks: hence to explain the physical substrate of perception, one must look at dynamically unfolding interactions, rather than 'object'-like structures in the brain.

This kind of thinking, that is, the idea of the temporality of experiences, is embedded in enactivism; accepting enactivism means accepting the temporal, dynamic nature of experiences. Sensorimotor action is not conceivable without temporality. The same applies to ECM. If experience is extended at all, the extension cannot be restricted only to the material sphere, for the temporal dimension follows necessarily. As Noë (2006b, 32) writes: "When you perceive an event unfolding, it is not as if you occupy a dimensionless point of observation". Temporality is a necessary element for interchange between the bodily and environmental elements and cannot be overlooked when explaining their constitutive basis.

This dynamic unfolding in time is close to how we seem to acquire experiences. And in the case of experiences, which are inherently subjective first-person phenomena, to dismiss the *what-it-is-like* side is to dismiss

[10] A further and unbiased theory regarding the matter at hand is *dynamical systems theory*, which is a mathematical theory of properties of abstract dynamical systems. In cognitive sciences, it has been used as an alternative to computationalism. According to this theory, cognitive systems resemble dynamical systems rather than computational systems (see, e.g. Van Gelder 1998).

the referent of the concept. However, not everybody agrees with the ubiquitous temporality of experiences. For example, Crick and Koch (2003) claim that we do not experience motion, but only static snapshots, and the alleged experience of motion is only imaginary, "painted" on snapshots. Consequently, there is an ongoing debate about how we experience temporal phenomena (see, e.g. Dainton 2008; Arstila 2016; Kiverstein and Arstila 2013). Without diving too deeply into the questions of philosophy of time consciousness, let us briefly compare different ways of explaining the alleged temporality of experiences. This will help us see why we wish to embrace an *extensionalist* view of time experience.

Theories of time consciousness can be divided based on whether they are committed to dynamical or static (or diachronic vs. synchronic) explanations of experiencing time. Further, the explanandum is at two levels—material and experiential—and they might have a different explanans.[11] According to *extensional models*, not only the contents of experiences but also the vehicles (sub-personal realisers) of those experiences are dynamical. The commitment to dynamic, temporally extended vehicles is what differentiates extensional models from *retentional models*, which accept the dynamic nature of the contents of experiences but do not require that their material vehicles are dynamical (whether the retentionalists hold that they are necessarily instantaneous remains unclear). Finally, *cinematic models* hold that both the contents and vehicles are instantaneous—the apparent temporality is only an illusion, as in displayed films, which consist of rapid sequences of instantaneous still images or snapshots. (The terminology is from Dainton 2018.)

The 4E-views are clearly at odds with the cinematic view, which is committed to the view that we never experience movement, change, and succession, for those are considered illusory. All we perceive is detached static time-slices; direct non-inferential experience of those temporal phenomena is not possible. For example, Crick and Koch (2003) claim that the alleged experience of motion is only imaginary, "painted" on snapshots. This view fits well with the "grand illusion" view described in the previous chapter. We think we enjoy perceptual experiences of the detailed scenery around us with moving objects in it, but we are mistaken—it is only an illusion.

[11] Although some theories of philosophy of time consciousness deny this difference in levels, for example, a view where time is treated as "its own representation" (see the discussion in Kiverstein and Arstila 2013, 447).

An obvious argument against the cinematic model is that it cannot explain observable dynamicity, and it is hard to see why an experience should be explained in a way that contradicts that experience. It seems obvious that we are capable of experiencing movement, change, and succession; otherwise we could have never had an experience of an approaching car or a ball, and it would seem very odd to claim that we haven't. As embodied creatures, we survive by avoiding certain moving objects. A series of instantaneous slices cannot be sufficient in order to experience movement. While cognition and perception can occur at least partly unconsciously, experience, by definition, cannot. (Another question is the reportability of conscious experiences: it is not straightforward how to interpret reports of experiences in empirical settings.)

However, this argument is not enough to refute the retentional model (e.g. Husserl 1991), since retentional models admit that at the personal level, experiences *are* temporally extended. This makes the retentional model a far better candidate, but it is still wanting at the vehicular level. To settle for the retentional model would be the same as to accept sensorimotor enactivism but refute ECM, that is, to accept situatedness at the personal level, but not at the sub-personal level. For reasons that were presented in Chap. 2 (see Sect. 2.3.6.), this position is not tenable—the personal level determines the sub-personal level.

Furthermore, brain events themselves take time. Neural processes are *processes*, never instantaneous brain "states", for example, action potential takes time. Clark (2009, 978) acknowledges this in his otherwise critical paper: "for information to become conscious, some amount of time needs to pass, so that normally there is no way the brain can 'in an instant' reach the kind of state that supports conscious experience". As a consequence, some kind of temporality at the neural level is already agreed by all parties. However, the retentionalists think that the temporality at the personal level is not in accordance with the temporality on the neural level, and therefore the temporality at the material (neural) level is not important for settling this issue. The problem of the retentionalist model is exactly this: according to this model, the phenomenal and material events do not match. Hence, neither the retentional model nor the cinematic model is up to the task—what we want is an extensionalist model.

Even though "hearing a melody" is an often-used metaphor for describing a temporal phenomenon, in this case it is slightly misleading, since my claim goes to an even more elemental level. Not only a melody (whose temporality nobody would deny) but also a single note has a temporal character. According to Noë (2006b), hearing a note necessarily involves a temporally extended coupling-relationship between the hearer and her environment. According to Foster (1991, 247), we couldn't even conceive of an auditory experience without the temporal dimension. In the same way as there couldn't be an experience of colour without the region of what the colour pervades, there couldn't be sound experience without the period of time that the experience of sound fills.

The same applies for tactile modality. It is quite evident that tactile experience is very closely connected with movement, and movement is necessarily a temporal phenomenon. If we want to know what some surface feels like, we need to move our hands (or the tool we are touching it with). Without movement, which necessarily includes the temporal element, there is no way we can have tactile information about it. Noë (2004, 96–100) refers to Bishop Berkeley, who held that touch is the only intrinsically active and spatial sense; it is in fact "a kind of movement". (Noë himself thinks, of course, that even though Berkeley is right about the tactile sense, he is mistaken in that he says it is the *only* active modality.) Noë also refers to O'Shaughnessy (2000, 658), who writes that "touch is in a certain respect the most important and certainly the most primordial of the senses [...] it is scarcely to be distinguished from the having of a body that can act in physical space". This underlines how touching an object depends on its resistance when we contact it by pushing, poking, nudging, stroking, and so on. The tactile experience is dependent on the very act of movement.

Thus, it seems clear that auditory and tactile experiences are temporally extended phenomena. However, perhaps the temporality of visual modality is not as clear-cut a case because of the intuitive feeling that when we see say a red object, we are passive observers rather than active skilful probers. As presented in the previous chapter, the sensorimotor theory of perception (O'Regan and Noë 2001; Noë 2004) is an example of an all-encompassing dynamic view that argues that all perceptual experiences (including sensations, e.g. the sensation of pain) are dynamic and based on our (temporal) actions—what we *do, probe, and inquire*, and not on static states that we passively *receive*. Hurley's (1998) criticism of the "input–output model" is another example that challenges the view that we are just passive receivers of phenomenal states.

The experiences ECM deals with, and which are in need of constitutive explanation for ECM to be a solid theory, are perceptual experiences related to sensorimotor action and movement, especially of tactile and visual modalities. Examples include moving around in a room with the help of a stick, feeling with tactile-to-tactile sensory substitution technology, or seeing with visual-to-tactile or visual-to-auditory sensory substitution technology. The *what-it-is-like* side is present in them too (as in all experiences), but they are also action oriented and goal directed in the minimal sense that an active subject is in interaction with her environment. Hence, those experiences are very much dependent on real-time interchanges between the (embodied) subject and its environment.

Even if we were internalists about the material realisers of experience, the notion of experience would still require temporality. An experience of, say, sound or touch is inconceivable without the event in time it pervades, no matter whether it is realised in the brain alone or whether environmental interactions are taken on board. Hence, the dynamical–temporal view is not tied to ECM alone.

Moreover, the static vs. dynamic question is not only related to the internalism–externalism debate. The dynamic view has gained support from other areas of philosophy, too. One of the best-known "dynamicists", Tim Van Gelder (1998, 621) has given a characterisation of the dynamical stance on cognition and contrasted it with the computationalist view as follows. Dynamicists are interested in how things change; computationalists are interested in states—state by state. Dynamicists understand a system's states in terms of its position with other states; computationalists focus on internal structure, what the state is made of. Dynamicists think that cognition is laid out temporally, like "speech as opposed to the written word"; computationalists think that everything is laid out at one moment of time. This is of course a simplified picture of a very complicated topic but gives a general idea about the differences in these two approaches.[12]

[12] Another strand that stresses dynamicity and comes outside the externalist views in philosophy of mind is *process philosophy*. The label refers to a variety of views held in, for example, phenomenology and pragmatism that appeal to the role of dynamic processes and temporal activities rather than timeless, static substances when forming accounts of being and reality (Rescher 2000).

To recapitulate, the need for dynamic explanation stems from the fact that the experiences ECM deals with are based on reciprocal interaction between the subject and her environment—and interaction is always inherently a temporal and dynamical phenomenon. The extension that we are dealing with in ECM is not only spatial but spatiotemporal. Accordingly, the notion of constitution needs to be such that it takes into account the dynamic nature of experiences. In the next section, I will move from the explanandum to the notion of constitution.

3.2 Defining Constitution

Constitution is a *building relation*: the whole is made of its parts (Bennett 2011). It is usually explained as a synchronic relation between higher- (*constituted*) and lower- (*constituent/s*) level phenomena, although the synchronicity is not a necessary feature. Causation, on the other hand, aims to give *aetiological* explanations that explain events: their origin, changes in their properties, and so on (Ylikoski 2013). Causal explanation explains what *caused* a phenomenon or event, and constitutive explanation explains *what* the phenomenon *is*, what it *is made of*. There is no reason to assume *a priori* that constitution should be something internal and causation external (to the organismic boundaries).[13] In what follows, I will present three different ways to outline the causal–constitution distinction: *material*, *mechanist*, and *diachronic–dynamical*. I will compare how they fit with EM and ECM. The three models might all serve for some purposes, and my aim is not to judge them for being unable to explain *any* constitutive relation—I only examine the cases as they bear on EM and ECM.

3.2.1 Material Constitution

The "standard" understanding of constitution derives from analytical metaphysics: it is a static view of *material constitution*. It is also the (often

[13] Even though this assumption is rather common. Hurley (2010, 106) calls it the "causal–constitutive error *error*". This kind of assumption is also related to a "primitive notion of spatial inclusion". Kaiser (2018) shows with examples from biology how spatial inclusion is neither sufficient nor necessary for something to be a constitutive part of a biological entity. She goes on to argue that the spatial inclusion criterion is also circular/trivial because demarcating the spatial boundary of a whole by positing all the parts in it already presupposes what its parts are.

unconscious) presupposition of the writers who accuse EM of having committed the C–C fallacy.

> Material constitution can be summarized as a *synchronic* one-one relation between spatially and materially coincident objects of different kinds, or as a many-one relation, where one object or entity is constituted by an aggregate of objects or entities. (Kirchhoff 2013, 320)

In the standard material understanding, constitution is considered a material realisation of an object. The basic example is that Michelangelo's *David* is materially constituted by the marble it is made of—it is nothing over and above that piece of marble (Wasserman 2004; Wiggins 1968; for critical a view, see Kirchhoff 2013). A material reading of constitution might explain the constitution in material objects like statues or tables and chairs. I am not taking any stand on whether it is a useful theory for explaining the metaphysics of spatial objects. However, what matters is that it is not a profitable way to explain constitution when discussing EM, ECM, or philosophy of mind in general.[14]

As Kirchhoff (2013, 321) analyses, David does not unfold in time, but it is constituted at a certain moment of time *t*. Temporality is not essential for its constitution relation. Kirchhoff continues that David's case reveals another feature of material constitution: the constitutive relation that holds between the marble and David coincides materially and spatially. However, demarcating the constitutive relation based on sharing the same material parts is problematic. As Kaiser (2018, 8; see also footnote 73 in the previous section) shows, the "spatial inclusion" criterion faces trouble. It assumes *a priori* "natural" boundaries based on what the spatial inclusion criterion already presupposes. However, this criticism is targeting the life sciences and may not apply to simple cases of material constitution, such as the statue of David.

Even though material constitution might explain some phenomena, the crucial point (and the only statement I am making about it) is that it cannot explain the phenomenon under scrutiny in my research because material constitution is unable to explain non-static phenomena. It serves as the basis for an atemporal type of explanation in the sense that constituted

[14] It might be the case that many defenders of EM and ECM also assume a concept of constitution along these lines because many of them have not delved into analysing the notion of constitution.

entities are assumed to be such that they do not unfold in time but are explained only from given moments—in specific instants of time *t*, snapshot by snapshot. Temporal extendedness plays no role in constituting the entities under examination. Because of its static and atemporal nature, the material conception of constitution is unable to explain cognitive processes and (especially) dynamic conscious experiences because of their innermost dynamical and processual nature.

Occurrent cognitive and experiential states need a different kind of concept of constitution than that which applies to material objects because what is essential to the former is not the static one-to-one relation but a temporally extended dynamic unfolding. So, if the material view of constitution cannot be used to explain cognition, let alone occurrent phenomenal experiences, what are the alternatives? As better candidates for explaining constitution in mental processes, I will consider the ***mechanist*** and the ***diachronic–dynamical*** conceptions of constitution. They are both better suited to accounting for the processual and dynamical nature of mental events. Moreover, they make better sense of not only extended mental processes but of all kinds of temporally extended dynamical phenomena.

One might oppose this line of proceeding because it looks like the result depends on which reading of constitution we choose. If we apply a static view of constitution, the relation of external elements and mental phenomenon is not constitutive. If we apply a non-static view, the relation is constitutive. A critique might ask what allows us to disregard the static view and instead apply a non-static view. However, I think we are entitled to discard the material view because it cannot explain dynamical processes like perceptual experiences—whether they are extended or not. The problem is that the explanans does not accord with the explanandum. The choice of the theory of constitution is thus independent of my argument for ECM.

To recap, it is not ECM as such that needs a different kind of understanding of constitution, but the whole dynamical view about the mind that is defended by all the E-views. As I have stressed, the problem with the material view is that it is built on a static understanding of constitution. Externalist views, alongside all dynamic views (which I think all the accounts about conscious experience should be for them to be credible theories), need a different approach that appreciates the dynamic nature of the phenomenon to be explained. The solution I start with derives from the ***new mechanist approach*** in philosophy of science.

3.2.2 Mechanist Constitution

In the *New Mechanist Approach* (e.g. Craver 2007a; Craver and Bechtel 2007; Bechtel and Abrahamsen 2005), a phenomenon is explained by referring to its underlying mechanism. The *explanandum* is the phenomenon, and the *explanans* is the mechanism (Craver 2007a, 139). Constitution is an inter-level relation between higher and lower systemic levels. The methodology behind this approach is *naturalistic*, and is committed to *descriptive adequacy*, where a theory needs to describe an actual scientific practice in order to be an adequate theory (Craver 2007a, 19; Kaiser and Krickel 2017, 750). The mechanist view is very different from the static view of constitution in analytical metaphysics: it aims to *explain* phenomena within empirical research. It is a kind of scientific explanation rather than a metaphysical enquiry into the ontological nature of material objects. Hence, the notion of constitution cannot be used in the same way as it is used in metaphysics, or in the static material view discussed above.

It is understandable that many EM supporters have been tempted to use the mechanist approach as a backup theory for EM, since it promises to get access to scientific explanations. It is likely that the "constitution-turn" in the EM literature exists mainly because of the mechanist view; without the considerations in philosophy of science, probably philosophy of mind and the cognitive sciences would not have paid so much attention to the causal–constitution question in the first place. However, the mechanist view has mainly been used to support EM, and only recently there has been the first attempts to reconcile it with ECM. The most important of them is Kirchhoff and Kiverstein's (2019) monograph that builds on Kirchhoff's earlier accounts of diachronic constitution (that will be discussed later in this chapter). I also think Gallagher's (2018) account is compatible with ECM, although he may not have intended so, because he has stayed agnostic about ECM elsewhere.

The mechanist view (at least its orthodox reading) makes a strong divide between aetiological (i.e. causal) and constitutive explanation: constitutive relation is a non-causal relation and vice versa. Causal mechanisms cause phenomena, whereas constitutive mechanisms non-causally ground, build, or constitute phenomena—and the theory promises to reveal where the causal relation ends and the constitutive begins. In the mechanist view, constitutive (and also causal, but causal mechanisms are not under scrutiny in here) relation holds between the mechanism and the phenomenon. Importantly for my purposes, the mechanism may consist of different elements: not only material objects, but also actions, behaviour, or interactive loops between the agent and the environment.[15]

According to Craver (2007a), a phenomenon is realised by its mechanism in the following way. Lower-level components of the mechanism constitute the higher-level phenomenon. The components are causally related to each other, and they constitute the system (S), and the higher-order process (ψ-ing). The *intra-level* relations are causal, and the *inter-level* relations constitutive. Something is a compositional constituent of a system if it is a working part "that non-causally results in the 'work' done by the relevant whole" (Gillett 2013, 319). Constitutive elements are *part of the phenomenon* itself; in the constitutive relation, the parts and how they are organised constitute the phenomenon (the whole). It is a compositional view: the phenomena under examination consist of parts that together constitute the whole.

If we consider Craver's model in terms of EM, S ψ-ing (the whole) could be, for example, solving a mathematical problem, and $x1\varphi1$-ing–$x4\varphi4$-ing (the parts) sub-personal processes in the following way: "say, $x3\phi3$-ing would be the brain processing information [...] $x1\varphi1$-ing might be the eye performing a saccade, where $x4\varphi4$-ing might be writing with a pencil" (Aizawa 2014, 24; see also Gallagher 2018). There is causal interaction between the lower-level parts (the brain processes, saccade, and writing with a pencil), and they together constitute the higher-level process (solving a maths problem).

The following three features describe constitutive relations as they are usually understood in the mechanist framework. Firstly, the constitutive relation is *asymmetric*: the components constitute the whole, but not the other way around, and the parts are independent of the whole, but not the other way around. Secondly, it is *irreflexive*: nothing constitutes itself. Thirdly, it is *synchronous*: constitution takes no time. It is common (but there is no general agreement) amongst the proponents of the mechanist view to claim that causal processes take time, but constitution does not—and hence processes are out of reach of constitutive explanation (Ylikoski 2013, 282.) This last feature is problematic, as we will soon see.

[15] But this issue is not straightforward within the proponents of the mechanist view. The mechanist view quite explicitly (e.g. Craver 2007a; Ylikoski 2013) requires that the constituents and constituted spatiotemporally overlap—hence it is difficult to see how "actions, behaviour, or interactive loops" could fit the picture. This touches on the ambivalence at the very heart of the mechanist view (I will return to the topic later in this section).

According to the mechanist view, what differentiates constitution from causation is that causal relations are not synchronous but are diachronic by nature. Thus, they differ in the third feature. All parties agree about the diachronicity of causal relations, but the necessity of the synchronicity of constitutive relations will be under critical scrutiny soon (even though it is widely accepted by the advocates of the mechanist and material views). The third feature (synchronicity) makes the mechanist model problematic for explaining the constitutive relations in mental processes. Since constitution does not take time according to the mechanist picture, whereas mental processes do take time, they are clearly incompatible.

The incompatibility between the view I am supporting and the mechanist view can be found in a passage from Petri Ylikoski (2013, 280). While he agrees that the ultimate goal of constitutive explaining is to understand a system's *behaviour*, he nevertheless claims that this can be achieved while leaving out some behavioural aspects and "orchestrated activities of the parts". He admits that constitutive explaining makes "heavy abstractions", but that they are needed for picking out the constitutive questions:

> When building an understanding of a system's behavior, scientists pose more focused questions that *abstract away some dimensions of the system's behavior*. [...] One could say that the question only addresses a synchronous time-slice of the system. (Ylikoski 2013, 280, italics added)

This might be a useful way to deal with the examples philosophers of science use (e.g. questions in philosophy of biology). However, it simply does not work with the phenomenon I am working with. It is impossible to abstract away the intrinsic dynamical–temporal aspects from experiences—if we did, what is left over would be something else than experiences.

In spite of these inadequacies, the mechanist view is a much better way to understand constitution if we compare it with the material view because it relies less on metaphysical assumptions and instead aims to grasp how natural scientists explain phenomena. This promise of scientific adequacy, or arguably empirical adequacy, is undoubtedly one of the main reasons why some advocates of EM have used the mechanist view as a supporting theory.

So, how does all this relate to EM and ECM? It looks like the mechanist view *can* explain constitution in non-occurrent dispositional beliefs (like Otto's case) because they are not temporally extended in nature. However, the mechanist view cannot explain constitutive relations in ECM or in occurrent cases of EM. This only applies if we interpret the mechanist view in its orthodox reading. Nevertheless, it would also explain occurrent EM and ECM if understood in a less orthodox way that takes diachronic relations on board—but then it could no longer be called the mechanist view. The problem with the mechanist view (for my purposes) is that it is synchronic/contemporaneous. Instead, we need a model of constitution that takes into account the intrinsic dynamic and diachronic nature that experiential states bear.

Thus, the mechanist view of constitution only fits with a narrow understanding of EM (that considers non-occurrent cognitive states). However, the mechanist model lacks the ability to explain experiences—and also occurrent cognitive processes—because of their highly dynamical, temporal, and process-like nature. That is why we need to combine the aforementioned features with our concept of constitution.

The Ambiguity in the Mechanistic Model

Apart from its very explicitly stated requirement for synchronicity,[16] the mechanist view would be a good way to explain the constitutive relation in ECM and EM. The reason for this is that whether or not the proponents of the mechanist view so intended, it would seem that diachronicity is already in their notion of constitution, even though they state in various places that the relation is synchronic. An example of this kind of ambiguity can be found in Craver's (2007a) *mutual manipulability* criterion (which is probably the best-known application of the mechanist theory). It relies on *causal* relations, defined in terms of Woodward's (2003) interventionist theory.[17]

Some mechanists explicitly stress the importance of processes (e.g. Kaiser 2018; Krickel 2018; Kaiser and Krickel 2017; Harbecke 2010). Some other mechanists are committed to views that take diachronicity on board, whether they intended to or not. For example, it has been said of Craver's (2007a) analysis of *action potential* (a change in electrical potential in neurons, also known as nerve impulse) that it requires a commitment to temporally evolving phenomena. Kaiser and Krickel (2017,

[16] For example, Craver (2007a, 153) leaves no doubt about the matter: "the constitution relationship is synchronic". Ylikoski (2013, 282) agrees: "the relation of constitution is synchronous".

[17] All of these topics are covered in more detail in the following section.

755–756) show that the action potential cannot be explained only by referring to spatial parthood, but that temporal parthood is also needed. It is what the action potential mechanism does and when it does it. An ion-channel opening must occur during an occurrence of the action potential for it to be constitutively relevant for it. Marie Kaiser and Beate Krickel go on to argue that (also in other cases besides the action potential) mechanisms are composed of objects and occurrents,[18] or rather they are *object-involving occurrents* (p. 768).

> [M]echanisms themselves not only have locations, more or less clear spatial boundaries, and spatial parts, they also have durations, involve changes, and have temporal parts. Hence, mechanisms are also things that occur. There is something going on when a mechanism occurs. Mechanisms do not merely consist of spatially extended component-objects that are only disposed to act but do not actually perform any action. (Kaiser and Krickel 2017, 765)

Thus, the mechanist view comprises slightly different readings, and it seems like, for example, Kaiser and Krickel's view that mechanisms can contain external and temporal parts fits rather well with my purposes. However, even if we agreed that diachronicity was already (more or less implicitly) embedded in the mechanist model, I couldn't call the view I am supporting "the mechanist view with a diachronic twist", since the orthodox mechanist view so clearly commits itself to synchronicity, and I myself so clearly reject it.[19]

Furthermore, Craver is not an internalist: he acknowledges that the mechanism often extends outside the organismic boundaries. In the case of the action potential, for example, ions that are outside the neuron are still part of the mechanism. "The mechanism of the action potential relies crucially on the fact that some components of the mechanism are inside the membrane and some are outside" (Craver 2007a, 141). Interestingly, Craver then proceeds to cognitive extension, stating that "many cognitive

[18] Their view has similarities with *entity–activity dualism* as a characterisation of a mechanism, according to which mechanisms consist of both entities (as in material objects) and activities (what objects do) (e.g. Machamer et al. 2000). It is also similar to Craver's (2007a, 189) notion of *acting entity*, which also stresses what objects do.

[19] Whether the mechanist view is already diachronic (and whether the two views should be separated at all) is an interesting question and would no doubt deserve to be examined in depth. However, since my thesis about ECM is not dependent on this question, I will not go into more details about the topic.

mechanisms draw upon resources outside of the brain and outside of the body to such an extent that it may not be fruitful to see the skin, or the surface of the CNS, as a useful boundary" (Craver 2007a, 141).

Hence, it seems that Craver himself is happy to extend cognitive phenomena within his mechanist model, or at least he does not oppose the idea. However, the problems I raised earlier remain. The request for synchronicity found in the heart of the mechanist model stands in the way of extending (most) cognitive phenomena, and all experiential phenomena.

Why is the restriction of the diachronic side (and temporality) to aetiological relations so crucial for mechanists? It is important to notice that synchronicity is not a necessary property of constitution. Even some mechanists admit that "we should not require that building relations be synchronic" (Bennett 2011, 94). It is embedded in the orthodox reading of the mechanist view, but we can have a consistent view that drops this requirement while keeping the distinction between aetiological and constitutive relations.

3.2.3 *Diachronic–Dynamical Constitution*

Kirchhoff (2013, 341) has aptly phrased the problem that arises from the mechanist view:

> [H]ow does something that is inherently temporal (the complex causal relations between processes and their component parts at a lower level) atemporally constitute something that is inherently temporal at a higher level (e.g., the distributed process of remembering)?

As hinted in the previous section, I go along with Kirchhoff's worry and do not think this is a possible or the most plausible way to explain the constitutive relation. It is difficult to see why the diachronic lower level and the diachronic higher level should be connected with a non-diachronic relation.

The third view of constitution to be presented (and the one I argue for) responds to this problem and casts aside the requirement for synchronicity. I shall call it the *diachronic–dynamical view of constitution*. This move allows there to be causal (or causal-like) features within the notion of constitution. It stresses the dynamic nature whose roots were already set down in the mechanist view (at least implicitly, and even explicitly in some texts, e.g. Krickel 2018; Kaiser and Krickel 2017). Thus, the central features of this view are that it is both dynamic and diachronic. The notions of causal and constitutive remain separate, but this view also allows there to be non-synchronic, non-contemporaneous elements in the constitutive relations.

The static notion of constitution (and its use with regard to EM) has recently been criticised from an enactivist viewpoint. Kirchhoff (2013, 2015) has developed an alternative (to the mechanist) view of constitution that he calls "diachronic process constitution". Gallagher (2018) has been working with a quite similar dynamical view of constitution using Varela's (1999) neurophenomenological analysis of cognitive processes occurring on different timescales. My formulation of diachronic–dynamical constitution draws from both of these accounts.

If the paradigm example of material constitution was the statue of David, with diachronic–dynamical constitution, it could be a Watt (centrifugal) governor. This example has been used in cognitive sciences to describe a dynamical system in general (e.g. van Gelder 1995). Kirchhoff (2013, 328–329; 346), however, uses the Watt governor as an example precisely for diachronic constitution. A Watt governor is a mechanism that controls the speed in a steam engine so that it maintains constant speed for the flywheel. It contains two "arms" that are free to rise and fall in accordance with the speed of the governor on account of centrifugal force. As a result, the governor controls the amount of steam that drives the flywheel.

> The key feature of the governor's behaviour is the angle at which the arms are hanging, for this angle determines how much the throttle valve is opened or closed. Therefore, in order to understand the behaviour of the governor, we need to understand the basic principles governing how arm angle changes over time. (van Gelder 1995, 356)

In the governor, maintaining the constant speed of the flywheel does not happen at a certain time *t* but is rather expanded in the unfolding process. Hence, it is a temporally extended system. Moreover, the Watt governor is not tied to a particular location: "It has parts, to be sure, and its overall behaviour is the direct result of the organised interaction of those parts" (van Gelder 1995, 355). Hence, the processes it is performing "must be individuated in terms of their roles in a dynamic context" (Kirchhoff 2013, 346). Those processes are what constitute it.

Regarding ECM and the notion of constitution, the case of the Watt governor shows that processes are not tied to a single location necessarily, and not to a single moment of time. The C–C fallacy objection treats its target with the concept of constitution drawn from the statue of David rather than the Watt governor, and that is why it seems to succeed. However, experience cannot be treated in these terms, and hence the C–C fallacy objection's outcome is misleading.

Accepting causal features for our concept of constitution doesn't mean we should rule out compositional relations, that is, the diachronic–dynamical view does not require non-synchronicity, but rather that constitutive relations can include both synchronic and non-synchronic properties. The big picture remains similar to the mechanist view: constitution is an *inter-level* relation, and causation is an *intra-level* relation. As Kirchhoff (2013, 324) writes, constitutive relations occur "between the putative higher-level distributed cognitive processes and its lower level sub-processes and their components".[20] What is different from the mechanist view is that in the diachronic–dynamical view, the constituents also include "sub-*processes*". Accepting processes as part of the constitutive relation brings temporality on board because processes are necessarily temporal.

It is quite natural to say that processes occur or unfold in *time*. But what does it mean to "unfold in time"? One should notice that processes are continuous in time on both levels: both *constituted* (at the higher level) and its *constituents* (at the lower level) are temporal (in their own way). Kirchhoff (2013, 234) describes the constitution relation as "dynamically embedded at multiple timescales". There are multiple timescales included, and they intertwine in a complex way; for example, neural processes, sensorimotor actions, and social practices lie on different timescales. Temporality is not restricted only to intra-level relations. Higher-level phenomena (e.g. solving a maths problem or hearing a melody) usually stretch over a longer period of time than lower-level components (e.g. neural and extra-neural processes, such as the manipulation of pen and

[20] Kirchhoff's use of the term 'component' is slightly misleading since the term can (and should, in my opinion) refer to any kind of constituents—including processes, even though in his use these two are separated, and it sounds like he refers to material parts when talking about components (at least in the sentence quoted). My use of 'component' is wider than Kirchhoff's: it includes not only material parts but also (sub-)processes. I don't see as very useful the task of trying to carve and separate the material parts (components, in Kirchhoff's use) out of sub-processes.

paper). Thus, it is impossible to fix the constitution relation synchronically, since the phenomenon to be explained is temporal on all levels.

So far, we have mostly discussed the diachronic side of the diachronic–dynamical view. To get to grips with the second part or the term, we need to look closer at how the concept of diachronicity is used in a *dynamical* sense. Getting rid of the snapshot picture is very important since snapshots cannot constitute dynamic processes. However, we need more than this. Diachronicity should not be understood as a sequential, linear process. As was stated, the diachronic–dynamical view stresses the process-like nature of evolving experiential events as opposed to synchronic contemporaneous states. However, the processual nature of experiences doesn't need to mean that the relation between constituent sub-processes is sequential, that is, linear in time. Experiential processes are temporal not only in the sense of succession in time, but they are also *non-linearly dynamic* across timescales. This is what I mean by the 'diachronic–*dynamical*' view. Perceptual experiences are not contemporaneous, but extending in time; not linear, but non-linear; not static, but dynamic.

Let us look back on the levels familiar from the mechanist view (see Craver's picture in Sect. 3.2.2.). The lower level consists of components that are tied together with causal relations (an advocate of the mechanist view would agree so far). According to the diachronic–dynamical view, it is this mutually influencing causality, that is, *processuality*, between the components that *makes* them the constitutive components of the occurrent cognitive/conscious process. Hence, it is causal/diachronic relations that ground constitution. This kind of thinking can be found in Clark (2008) as well, even though he is not using the notion to explain the concept of constitution. His notion of *continuous reciprocal causality* fits very well with the diachronic–dynamical view: it is this continuous reciprocal causality that grounds constitution. (I will return to this notion later.)

The higher-level phenomena are temporal processes (e.g. counting with pen and paper), and their underpinnings unfold diachronically (e.g. brain processes, sensorimotor interactions with the hand and pen). The processual elements, that is, the "elements engaged in processes that constitute across different timescales the occurrent cognitive process", constitute the cognitive/conscious system (Gallagher 2018). Or to put it in other words, the processual relations among the elements (both external and internal) *are* the cognitive/conscious processing itself (part of it).

This way of thinking is familiar from enactivism. (However, enactivists have not written much about the causal–constitution distinction as such, and the question is not discussed in these terms.) In enactivism, it *is* the (causal) reciprocal relationships between the subject and the environment that constitute cognition—and according to enactivism, cognition includes experience. Cognition (and experience) *is* that interactive relationship. (For more details, see Chap. 1.) Hence, the diachronic–dynamical view fits best not only with ECM but also with enactivism.

To sum up, the diachronic–dynamical view of constitution is the best model for ECM because it takes into account the diachronicity (understood as wider than in a merely sequential sense) required in explaining the dynamical nature of experiences. Furthermore, of the three models of constitution I presented, it is the only one that can explain constitution in an enactivist framework—which is important since ECM relies on enactivism. The active cognitive system enactivism deals with must include not only synchronic relations but also diachronic processes, since the cognitive system and its processes are constituted on that precise reciprocal dynamic change that happens in time. The other two models of constitution are incommensurate with the phenomenon to be explained, whereas the diachronic–dynamical model is not.

Fair enough, a critical reader might say. Perhaps this understanding of constitution really is the most suitable for a phenomenon like conscious experience. But does that imply that it is true? I think this question is slightly misleading. It is enough to say that it is the best way to explain the phenomenon under scrutiny. For example, biological phenomena cannot be explained with the material view of constitution but require a view that is attuned to their intrinsic features (e.g. the mechanistic model of constitution). In the same manner experience—whether extended or not—requires an explanation that takes into account its special features. If they were left out, we would no longer be dealing with experience.

3.3 Demarcating the System Boundaries

We have now proceeded from the static view, through a mechanist view to the diachronic–dynamical view. In this section, the approach is more pragmatic: the aim is to test the view developed in the previous section and survey how it can be applied to demarcate what is constitutive and what is not. Constitutive relation is supposed to tell us which elements are genuinely components of the system and which are not. Giving an answer to "how does one demarcate the constitutive components from the causal background conditions of an experience?" will tell us not only what is a constitutive component of an experience but also the method to find that fact out. Therefore, by answering the causal–constitutive question, we will also get the other side of the coin: an answer to the problem how to demarcate the system boundaries. If we know the conditions for something to play a constitutive role, we also know that things that fulfil them are "inside the boundaries" of cognition/consciousness (where "inside" does not refer to the skin and skull barriers but to system barriers, which vary depending on the action we are examining). Therefore, those conditions function as general demarcation criteria for cognition/consciousness.

An eligible criterion for constitutive relevance will also respond to a criticism closely related to the C–C fallacy, that is, the "cognitive bloat" problem (Adams and Aizawa 2001, 57) or the problem of "rampant expansion" (Sprevak 2009). It is a worry that "cognition bleeds into everything". It is clearly an unwanted effect of the externalist theory if it allows the mind to extend for example, to the whole of the Internet (Clark and Chalmers 1998, 17) or to all the books on my bookshelf. Since the cognitive bloat problem is the outcome of committing the C–C fallacy, solving the C–C fallacy by giving apt criteria for constitution also solves the cognitive bloat problem. In other words, we need to solve the question of how to distinguish genuine components from causal background, and by doing so, we will also prevent the over-extension problem.

3.3.1 *Mutual Manipulability as a Demarcation Criterion*

Advocates of the externalist views have offered a wide variety of demarcation criteria. These conditions need to be fulfilled for there to be extension, that is, when we can count something to be a genuine constitutive component/part. The suggestions for requirements include, for example, Clark and Chalmers's (1998) original "glue & trust" conditions, "the responsibility condition" (e.g. Roberts 2012; Cash 2013), and Gallagher's (2013) appealing to enactive engagements that grounds the "ownership of a process". These criteria stem from the externalist views themselves, and they do not take into account the concept of constitution itself. (All these criteria are scrutinised in more detail in the following chapter.)

However, in what follows I will discuss a criterion that is neutral regarding EM and ECM, and instead starts from the notion of constitution itself. It provides a general criterion for demarcating what counts as a constitutive component, and I will discuss whether it can be used to demarcate the boundaries of cognition and consciousness in EM and ECM. This criterion is *mutual manipulability* (MM), and it derives from the new mechanist literature (e.g. Craver 2007a, b). Even though MM originates from the mechanist view, it can be used with the diachronic–dynamical view, too. (Of course, the diachronic–dynamical view itself rose from the mechanist view.) Whether MM works is not dependent on the differences between the mechanist and diachronic–dynamical views, but on other issues (which will be discussed later in this chapter).

Similar ideas have also been presented in the externalist side, for example, "continuous-reciprocal-coupling criteria" (Menary 2007; Palermos 2014; Kirchhoff 2016). Clark (2008, 24) himself calls *continuous reciprocal causation* the interaction that "occurs when some system S is both continuously affecting and simultaneously being affected by activity in some other system O". Clark uses it in the same way as MM is used—a criterion to test whether something is a real component—or rather a real coupled system in his terminology. For Clark, continuous reciprocal causation is, however, only a sufficient but not a necessary condition for extension. For him, only the glue & trust criteria are necessary (though they are not sufficient) conditions. Palermos (2014, 34) disagrees and takes the criterion of "continuous mutual interaction on the basis of feedback loops" to be a necessary and sufficient condition for constitution. MM, as I present it, is necessary and sufficient for constitutive relations (on condition that it is a valid criterion in the first place). This is how the mechanists usually interpret it (e.g. Craver 2007a, 159). Otherwise, it would be too easy to satisfy.

In MM, for something to count as a component (i.e. it being in a constitutive role for the phenomenon), an intervention on part(s) of the system (bottom-up) must affect the whole system, and an intervention on the whole (top-down) must affect its parts.

> [A] component is relevant to the behavior of a mechanism as a whole when one can wiggle the behavior of the whole by wiggling the behavior of the component and one can wiggle the behavior of the component by wiggling the behavior as a whole. The two are related as part and whole and they are mutually manipulable. (Craver 2007a, 153)

MM is supposed to be able to capture how (e.g. neuro-) scientists experimentally determine the system boundaries. It is an interventionist method of distinguishing components from the causal background. Inter-level impact is a way to settle the boundaries by testing whether intervening on the putative system components (parts) has an effect on the higher-level phenomenon (whole), and on the other hand, by testing whether intervening on the phenomenon affects the putative parts. Constitutive dependency relations are bidirectional, and the constitutive relevance can be tested based on this bidirectionality: whether the target is mutually manipulable. Causal relations are manipulative too but only in one direction (from cause to effect) (Craver 2007a).

There are at least three reasons supporting the use of MM as a demarcation criterion. First, it has no bias regarding the externalism–internalism debate, but it is a neutral way to set up the debate. Second, it is compatible with the interventionist view of causation (Woodward 2003), which is widely supported in philosophy of science these days. And third, I claim that it supports the diachronic–dynamical view (over the mechanist view, although it was originally part of the mechanist view).

Even though MM is part of the mechanist theory, it seems that it actually fits better with the diachronic–dynamical view. The reason is that in the mechanist view, causality and constitution are supposed to stay separated, but MM picks out constitutive relations that are causal, that is, inter-level wigglings are causal, so the innermost basis of constitution is actually causal. Craver uses Woodward's (2003) interventionist concept of causality to explain the inter-level wigglings. This Woodwardian interventionist method brings causality into constitution, which is problematic for Craver's original mechanist model (but not for the diachronic–dynamical view). Depending on the standpoint, this can be seen as a problem either for the mechanist view in general, or for Craver's original formulation of the mechanist view, or for the MM requirement, or for the causal–constitutive distinction in general. However, it is not a problem for the diachronic–dynamical view. MM, the Woodwardian interventionist model, and the diachronic–dynamical view fit all together—and can be used as a demarcation criterion when our scope is to set the boundaries of constitutive elements of experience in ECM.

Mutual Manipulability in Practice

Examples of how MM can be used to separate background conditions from constitutive parts stem from empirical experiments.[21] Kaplan (2012, 561) refers to Fisk et al. (1993), where they investigated the impact of gravitational forces on human sensorimotor capacities, such as reaching. They found out (quite unsurprisingly) that changing the gravitational force (from the normal background 1G) affects the performance of the sensorimotor task: the reaching performance does not remain the same in different force conditions. This is an instance of a bottom-up intervention: wiggling the putative component, which in this case is the gravitational force. So one side of MM is satisfied. According to Kaplan, if only the first part is obtained, we can take it as a preliminary but incomplete reason to consider the putative part as constitutive for the system.

To accomplish MM, top-down intervention must also have an effect on the putative part. An intervention on the subject in a reaching task, changing the set-up somehow—wiggling the whole—should have an effect on the gravitational force. This outcome is implausible enough, so we can rule out MM in this case. Hence, we can conclude that the 1G gravitational environment is not a constitutive component for the sensorimotor task but a background condition. In this same manner, we can also say that oxygen supply and blood circulation are not components of cognition because the two sides of MM are not fulfilled. Interfering with them bottom-up has a crucial effect on cognitive capacities, but interfering with cognitive capacities top-down doesn't affect the general abilities of respiration or circulation. So, with the help of MM, we can categorise them as background conditions for cognitive (or sensorimotor) capacities, not constitutive parts.

As our aim is to examine occurrent experiences, a relevant MM relation occurs between the putative components and the experiential state.

[21] Baumgartner and Wilutzky (2017) have criticised this central methodological claim of MM though. According to them, it is impossible to set the boundaries between causes and constituents empirically—instead the matter is purely *pragmatic*, for example, constitution outperforms causation in explanatory power. Their view is rather inconvenient for the devotees of the constitution-turn, and, for example, Kirchhoff's (2015) defence for opportunities of empirical research stemming from the causal–constitution debate (mentioned earlier in this chapter) is against this view. I think both sides are partly right: as the question itself is metaphysically loaded, it already sets the frame for possible outcomes. However, even if the framework turned out to give us only *pragmatic* results, the hypotheses can be used in further empirical investigations.

Gallagher (2018) puts it in more fine-grained words: the MM relation lies between the processual elements of the (putative) components and what he dubs *occurrent cognitive processes*. The latter captures the ongoing engagements of a system's cognitive process created by the processual elements of the components. Of course, instead of *occurrent cognitive processes*, my focus is on *occurrent experiential processes*, or *experiential processes*, for short. (Gallagher does not discuss the possibility that there would be any experiential properties involved.)

Intervening on the experiential processes (top-down intervention) affects the whole process. It can be done, for example, by setting a task and interfering somehow with its whole set-up. The aforementioned experiment of changing the gravitational forces during a reaching task could have been interpreted like this; however, it was not used as an outer intervention/wiggling in Kaplan's example, but as a putative component. (And as a putative component it satisfied the bottom-up wiggling side but not the top-down side—hence, Kaplan concluded that it is a background condition.) It is an interesting feature in MM how the two levels and interventions on them easily get mixed up. This could be due to the fact that actually they cannot be separated in the first place, or at least we cannot put our fingers on them, but all we get is holistic wigglings. (Baumgartner's *fat-handed intervention* tackles a closely related issue, but I will come to that in the next section.)

Anyway, for MM to be an eligible demarcation criterion for constitution, we must be able to differentiate not only between the two levels, but also the higher-level phenomenon (i.e. the experiential process) from the *intervention* on the higher-level phenomenon, and the lower-level components from the *intervention* on the lower-level components. Nevertheless, mixing up does happen in the literature relatively frequently. All this said, I think (at least to some extent) that the levels *can* be separated. Moreover, the interventions can also be separated from the phenomena and the components. For example, I would say that changing gravitational forces could be taken as an instance of a top-down *intervention* (against Kaplan's intuitions). This should not be conflated with the higher-level *phenomenon*, but as a means to test the phenomenon, for example, a reaching task.

Bottom-up intervention, on the other hand, is easier to track. It is targeted to the (putative) components and their processual elements, that is, how they are causally related to each other. This is easier to picture: it can contain wiggling the brain (e.g. with transcranial magnetic stimulation), wiggling some part of the body (e.g. tying hands together during a reaching task), or wiggling an external object (e.g. changing the way a calculator functions or change the material of a blind cane).

Let us consider the hallmark example of EM, poor old forgetful Otto. It should be noted that the part–whole relation in Otto's case is not between the notebook and the brain—these two are both (putative) lower-level components. The inter-level relation is instead between the notebook, or rather between the *use* of the notebook (the part), and the act of remembering (the whole). (Clark and Chalmers present it as a dispositional memory, though that is not necessary and not very useful for my purposes). It is clear that wiggling the notebook (bottom-up) has an effect on remembering, for example, tearing a page out makes a significant difference on the memory or the act of remembering.

But what about the other way around? Gallagher (2018) has suggested that wiggling Otto's remembering process (top-down) could be achieved by giving him the information in a different way. For example, Inga could tell him the address of MoMA, and that would make the *role* (i.e. the *use*) of the notebook redundant—and making the role of the notebook redundant counts as affecting it, at least according to Gallagher. Kaplan (2012, 567), on the other hand, suggests that in a top-down intervention, engaging Otto in a memory-demanding task would make him modify (by rewriting, etc.) the contents of his notebook when needed.

I think an even more simple top-down wiggling would be to intoxicate Otto and by so doing alter his remembering processes. Perhaps he would forget to consult his notebook or have problems in reading the words correctly—which would cause him to go to a wrong address. Hence, it seems that we can say that the Otto case fulfils both sides of the MM criteria. However, this outcome is rather limited. The general problem with the Otto case is that it is a highly speculative thought experiment and, as such, does not reveal much about any real cognitive capacities.

Still, for the sake of argument, let us look at how this example would work with ECM. In ECM, the whole is the occurrent conscious experience and the part some processual element of a putative component. Let us consider a case where a blind person is using a cane to move around in a room. The lower-level putative component is the cane and the process how it is used, and the higher-level phenomenon is the occurrent experiential process of moving around. It is clear that removing the cane affects the experiential process, but how could wiggling the occurrent moving experience affect the cane? But in fact, to satisfy MM, it doesn't have to affect the cane itself—but rather the *use* of the cane. Perhaps we could wiggle the moving experience by changing the way the blind person gets a sense of the room in such a way that the role of the cane will change: maybe it becomes redundant or it gives wrong or different results. If we add another external resource, for example, Inga or a guide dog, the cane becomes redundant. Alternatively, if we intoxicate the blind person, she might become unable to use her cane in the correct way anymore. It is important to remember that the intervention as such is not what we are examining; it is just a means to examine the phenomenon. In this case, it also looks like both sides of MM are fulfilled.

We might still want more empirical framing of the problem. Let us now consider an example that has actually been tested empirically: solving a maths problem. So far this experiment has been used exclusively regarding EM (e.g. Gallagher 2018; Kaplan 2012; Clark 2008), but I see no reason why it should not work with ECM as well (for reasons I will soon give).

When solving a simple maths problem, we use several faculties, which may include, for example, visual perception, language, memory, bodily positions and movements, pen and paper, calculator, and other devices. Intervening on the lower level could be done by intervening on neural processing (e.g. by transcranial magnetic stimulation or by some drug), or by preventing eye movements (e.g. saccadic eye movements as was done in Ballard et al. 1995), or by preventing bodily movements and gestures (e.g. if the hand gestures are prevented by forcing individuals to sit on their hands, solving a maths problem will take more time and be less accurate, as was shown in Goldin-Meadow et al. 2001.) (About the role of the gestures, see also the Chap. 2, Sect. 2.4.1.) These bottom-up interventions clearly have an effect on the experiential process—therefore the first half of the MM requirement is satisfied.

But what about the other way around, that is, wiggling top-down to detect the results in the lower-level components? Solving a maths problem is a complex set of cognitive and experiential factors, a complex experiential process. How can we get a grip on that as a whole? Wiggling the whole, as already hinted, can be achieved through somehow changing the whole set-up, for example, engaging the subject in a task and monitoring the lower-level effects. This is a method commonly used in neuroscience: engaging subjects in tasks and detecting changes in their neural activity. Gallagher (2018) contemplates ways of doing this in the case of maths problems.[22] He suggest we could wiggle the whole by setting up a time limit, or a competition to see who solves the task quickest, or having a distractive audience who witnesses the counting act.[23]

Gallagher himself does not consider it in the following sense, but I myself think it is clear that these kind of top-down interventions can have an effect on the components. They would probably change the bodily postures and gestures, eye saccades, use of external tools (e.g. pen and paper or calculator), brain processing, and so on. Therefore, the second part of MM is also satisfied, and we can conclude that the elements that were wiggled both ways are components of this cognitive (and potentially also experiential) process.

Even though others have used this example only in order to support EM, I think it is a hasty way to see the issue. People don't suddenly become Chalmers's zombies when solving maths problems—many cognitive tasks (but not all) are also experiential. As I have argued in the first and second chapters, there is often something it is like when undergoing a cognitive task. Of course, someone might say that this "something-it-is-like" is irrelevant for the cognitive task, that it is just some kind of added extra or epiphenomenal factor for the completion of the task. But is this view coherent? Clearly cognitive tasks (such as solving a maths problem) can involve experiential features (e.g. gesturing with your hands, counting using your fingers, playing with one's pen, smiling, and frowning). These features play a real role in how the task is completed (easier, quicker, more accurate), and they do feel like something. Several empirical experiments have shown that these features matter (e.g. sitting on your hands, preventing gesturing, preventing saccades, Clark and Chalmers's Tetris example),

[22] Gallagher's own train of thought goes in a wrong direction there, though. He seems to have mixed up the intervention and the target of the intervention, and infers from the lack of change in the intervention itself that it is not a component—which it was not supposed to be, but only a way to trigger the putative components.

[23] Perhaps an even more illuminating example of MM is the role of saccadic eye movements (Ballard et al. 1995; discussed in Kaplan 2012, 563–564; and in Clark 2008). It was shown experimentally that during a memory-demanding copying task, the subjects were three times slower when their saccadic movements were prevented (therefore the bottom-up side was satisfied). The task itself is the top-down intervention: the subjects needed to cope with the cognitively demanding task and for that reason begun the pattern with eye saccades. This example shows in a precise way how bodily skills are part of the cognitive functions. However, it speaks only for the embodiment thesis, not the extension thesis, and as such it is not straightforward evidence for my case.

so they cannot be dubbed epiphenomenal features of the cognitive task. Hence, we can say that this example speaks about both EM and ECM alike.

Limitations of Mutual Manipulability

The MM criterion for constitutive relevance has also been criticised, for various reasons. The most apparent problem that arose from the discussion above seems to be that it is still unclear whether we really reached the "whole" in the top-down "wiggling the whole". It appears that bottom-up intervention can be done, but top-down is hard to achieve. It might be difficult to put one's finger on the system as a whole. What exactly is the "whole" in the case of cognitive or conscious process? It is difficult to talk about the whole process without actually referring to some part of it. As Gallagher (2018, 219) has noted, the intervention "has to occur 'some where' or in some aspect" of the system—the whole might be a "part" too after all. It is difficult to find the right level of "wholeness" that we should be wiggling. This depends on how we frame the question in the first place. The differentiation between bottom-up and top-down interventions might be clear conceptually, but this might only be due to our theoretical framing of the question.

As we learned from the previous chapter, the connection between the sub-personal and personal levels is one of the most discussed topics in the history of philosophy of mind (although with different terminology). These two levels are connected with bottom-up and top-down methods, even though they do not have identical meanings.

On the one hand, using a bottom-up method (i.e. a sub-personal approach), we cannot obtain any whole, since there is no distinct whole in that level. The "whole" is just the combination of the material parts, and nothing more. It is not "the conscious experience being experienced" or "the subject of the experience" because that would be a personal-level phenomenon. All we get is a list of material parts, for example, brains, bodies, and blind persons' canes. On the other hand, using a top-down method (i.e. a personal-level approach), we cannot obtain any parts, because approaching from the phenomenological whole, the experience is not dividable into parts as it was at the sub-personal level. However, to have a constitutive relation, we need the relation between the constituent and the constituted. As I presented in Chap. 2 (see Sect. 2.3.6.), enactivism offers a solution and binds the two levels together.

Besides, the usefulness of MM for demarcating constitution would be doubtful if Kaplan (2012, 560) was right, and it only revealed a sufficient condition for something to be a component. If that was the case, one could not conclude from a failure to fulfil MM that something is not a component—it would be too favourable a criterion. However, as Baumgartner and Wilutzky (2017, 1122) remark, Kaplan's interpretation is wrong. They point out that Craver (2007a, 159) presents MM as a sufficient condition for constitutive relevance and a lack of MM as a sufficient condition for constitutive irrelevance. Therefore, MM is both sufficient and necessary.

In the literature, however, the main focus of the criticism towards MM has been the inconsistency between the requirement for non-causality on the one hand and the use of Woodwardian (causal) interventionism on the other (as mentioned in the previous sections). There have been various attempts to solve the problem (uphold the causal–constitution distinction while also upholding interventionism). One of these attempts is the so-called fat-handed intervention, where the intervention is said to wiggle both levels simultaneously by means of one common cause (Baumgartner and Gebhartner 2016; Baumgartner and Wilutzky 2017).

The reason for employing the notion of fat-handedness is the same worry that we are familiar with now. According to fat-handedness critiques, it is impossible to put our fingers (whether fat or not) on one level only—to *surgically* intervene on one level and thereby affect the other level. "Mechanist systems can only be manipulated on all their levels at the same time" (Baumgartner and Wilutzky 2017, 1113). However, Baumgartner and Wilutzky's aim is to criticise MM as a criterion for constitutive relevance in the case of EM—not to save it by using the fat-handedness argument.

Beate Krickel (2018) criticises the fat-handedness approach for several reasons, for example, that it loses the ability to distinguish between top-down and bottom-up interventions, since interventions always happen at both levels simultaneously. This is a problem because the motivation behind Craver's MM was precisely this: to show how scientists intervene ("wiggle") in their experiments using either bottom-up or top-down interventions. Her own solution (how to uphold both interventionism and the causal–constitution distinction) is based on a move that acknowledges the problems that arise from the requirement of synchronicity I pointed out earlier in this chapter. She interprets MM in diachronic rather than synchronic terms: the two sides that manipulate each other mutually are macro-level temporal parts of the phenomenon and micro-level component parts of the mechanism. Since the "whole" is a temporal process, an intervention on a lower-level component/process needs only to affect a temporal part of the whole.[24]

Even though my primary goal is not to solve the debates within mechanist literature about constitution, it seems fair to say that their understanding of the inter-level relations needs an update. Instead of synchronic, atemporal relations, they are better conceived as temporal, diachronic, and, hence, including causal features. This way of thinking saves interventionism from the fat-handedness arguments if we accept Krickel's (2018) account. It also enables us to keep the fundamental difference between causal and constitutive relevance.

Furthermore, MM leads to another problem: on what criteria should we decide whether MM is fulfilled or not? From what level of explanation should we go and look for the putative components? The level of explanation depends on what we want to explain, of course. The mechanist view is inherently a *compositional* view—and so is the diachronic–dynamical view as I have framed it here—the phenomena under examination consist of parts and wholes. If our aim is to explain *cognitive systems*, this kind of thinking makes sense: it is intelligible to talk about the parts and whole of a cognitive system. A cognitive *system* can be conceived as a whole that consists of its parts.

However, if the aim is to explain cognitive *processes*, the part–whole relation is not as straightforward. Gallagher (2018, 209) asks whether we should also think of cognitive processes as composed of parts, where one part of a process is distinct from another part of the process, and each part itself is a process, and together they form the larger process as a whole—without giving an explicit answer. Kirchhoff, on the other hand, explicitly states that in diachronic process constitution, processes have distinguishable individual parts. "A process might involve any number of component parts, but it always involves *some* parts" (Kirchhoff 2013, 343). The reason for this is that the explanation must stop somewhere. According to Kirchhoff, even though physics might tell us that, strictly speaking, individuals do not exist (because, e.g. only quantum fields exist), this level of

[24] There has also been a couple of attempts to formulate alternative demarcation criteria for constitutive relevance that do not rely on MM and interventionism. Harbecke (2010, 2015) and Couch (2011), for example, have argued for a *regularity theory of constitution* that rejects the idea that constitutive relevance can be explained by means of interventions.

explanation does not make much sense in cognitive sciences, where instead we must be able to identify not only entities but also their activities.

> To appropriately characterize "part" in processual terms, we must replace what Seibt has recently called the "particularist conception of individuals"—i.e., entities that are intrinsically individuated and which have a determinate unique location—with a view of individual parts that focuses not so much on "location but on '*specificity-in-functioning*' in the widest sense of 'functioning', i.e., focuses on the dynamic role of an entity (e.g., an activity) *within a certain dynamic context*". (Seibt 2009, p. 484; italics added, quoted in Kirchhoff 2013, 344–345)

With this in mind, it is possible to say that even experiential processes have parts—only without a determinate spacetime location. Moreover, I think there is even a simpler way out, and we can uphold a compositional view in the case of experiences. It makes sense to divide the material basis of experiences into parts on the sub-personal level or into lower-level components in MM terms. The explanandum of ECM is the material basis of perceptual experiences—answering the question which components take part in forming the material basis of certain perceptual experiences. Therefore, it is possible to interpret ECM in compositional terms, but it must be done along the lines of the diachronic–dynamical and not the synchronic approach. What is at stake here, however, is not the validity of ECM, but the validity of MM and whether it can be used to demonstrate the constitutive relation in ECM.

3.4 Conclusions

EM and ECM concern components and how they are organised, and are not concerned with the aetiological history of the components. The causal question would be 'What are the factors that cause cognitive/conscious processes, and what is their aetiological history?' Instead, EM and ECM are answers to the constitutive question 'What are the factors that make up the cognitive/conscious processes?'

The C–C fallacy objection does not work against either EM or ECM because its premise (the material view of constitution) is a wrong way to treat mental phenomena and especially dynamic experiences. The synchronic notion of constitution that is embedded in the C–C fallacy objection makes it unable to pick out the temporal dynamics that are inherent for the phenomena ECM is dealing with. Even though the C–C critique is the same for both EM and ECM, there is a difference in how they might answer the critique. This mistaken notion of constitution is in fact not inaccurate in the case of EM or at least with the original Otto example. The reason is that the Otto case deals with dispositional beliefs, and dispositional beliefs are in fact static in nature (at least if we compare them with experiential states). My belief about the dates of my parents' birthdays or about the address of a museum is not occurrent in nature as long as I don't consider them. But my experience of seeing a red tomato or feeling pain in my right foot is never dispositional but occur in a temporal mode, in a certain here and now. Phenomenal experiences are always occurrent in nature. Because of this difference, it is even easier to see why the C–C fallacy objection does not work against ECM than against EM.

However, in Clark's defence, we need to remember that Otto is not the only case of EM. The story of Otto serves rather as an intuition pump for a very much wider phenomenon. Clark's entire theory of EM avoids the problem that might arise if one focused only on dispositional beliefs (which could, at least on some reading, fit easier with the static view of constitution). Whereas Otto and his notebook represent an example of dispositional EM, for example, counting with pen and paper works as an example of occurrent EM. The latter example drifts away from a static understanding of constitution in the same way as cases of ECM.

The C–C fallacy criticism presented by Adams and Aizawa assumes that constitution is a synchronic, atemporal relation. As I showed in this chapter, occurrent cognitive and experiential states are temporal. Hence, the presumed synchronic notion of constitution is unable to pick out the phenomena it is trying to explain. The C–C fallacy objection relies on a false premise, and it is a question-begging argument.

Not only can we conclude that the C–C fallacy is mistaken but also that the diachronic–dynamical concept of constitution creates a positive way to understand the *extended*-relation as a constitutive one. We can conclude that some experiences are synchronically and diachronically constituted by external elements. I have argued in this chapter that both synchronic and diachronic relations can fall under the label of constitution—when all requirements are fulfilled, of course.[25] Furthermore, it has been argued that even if diachronic relations fell under causality, it would not justify the C–C

[25] The following chapter deals with these requirements in more detail.

fallacy criticism.[26] The reason for this is that the importance of external resources for our experiential events remains evident in any case—regardless of what our stance to the rather metaphysical causal–constitution quarrel is. For a cognitive scientist, their explanatory power is quite similar, and the ontological status doesn't make much difference (for similar views, see Sprevak 2010; Colombetti and Krueger 2015; Krueger 2018).[27]

However, it was important to show that ECM goes under the label 'extended' rather than 'embedded'. Another question is how fruitful that distinction is for an empirically attuned philosopher of mind or cognitive scientist. How could we ever test empirically whether something is causal or constitutive without just relying on our conceptualisations of them, which are more or less based on "one intuition over another" as Colombetti and Krueger (2015, 1159) have phrased it? The notion of constitution needed clarification, especially in the case of dynamical experiences. However, metaphysical inquiry will not provide us with answers about how experiences are actually extended and what their limits and criteria are. This is the more pragmatic and empirical task of the next chapter.

References

Adams, F. & Aizawa, K. (2001): "The Bounds of Cognition". *Philosophical Psychology*, 14(1): 43–64.

Adams, F. & Aizawa, K. (2010): "Defending the Bounds of Cognition". In Menary, R. (ed.), *The Extended Mind*. Cambridge, MA: MIT Press, 67–80.

Aizawa, K. (2010): "The coupling-constitution fallacy revisited". *Cognitive Systems Research* 11(4): 332–342.

Aizawa, K. (2014): "The enactivist revolution". *Avant* 5 (2): 19–42.

[26] For example, Levy (2011, 294) does not think our conceptualisation makes much difference: "If neuroethicists and philosophers wish to call only some part of human cognitive machinery our minds, and call the rest its [causal] scaffolding, so be it. [...] All that is lost [...] is the rhetorical power which comes from the identification of extended mechanisms with the mind".

[27] Ross and Ladyman (2010) go as far as to argue that the causal–constitution distinction plays no role in the natural sciences: therefore it should not be a problem in cognitive sciences either. However, I think they are wrong in drawing conclusions from the natural sciences to philosophy of mind. Although in physics, for example, the causal–constitution differentiation is quite meaningless, we cannot infer from this that this is necessarily so in all fields.

Arstila, V. (2016): "The Time of Experience and the Experience of Time". In B. Mölder et al. (eds.): *Philosophy and Psychology of Time.* Studies in Brain and Mind 9.

Ballard, D., Hayhoe, M., & Pelz, J. (1995): "Memory Representations in Natural Tasks". *Cognitive Neuroscience.* 7. 66–80.

Baumgartner, M., & Gebhartner, A. (2016): "Constitutive Relevance, Mutual Manipulability, and Fat-Handedness". *The British Journal for the Philosophy of Science* 67(3): 731–756.

Baumgartner, M., & Wilutzky, W. (2017): "Is it possible to experimentally determine the extension of cognition?" *Philosophical Psychology* 30(8): 1104–1125.

Bechtel, W., & Abrahamsen, A. (2005): "Explanation: A mechanist alternative". *Studies in History and Philosophy of Science.* Part C, 36(2), 421–441.

Bennett, K. (2011): "Construction area (no hard hat required)". *Philosophical Studies,* 154, 79–104.

Buhrmann, T., Di Paolo, E., & Barandiran, X. (2013): "A dynamical systems account of sensorimotor contingencies". *Frontiers in Psychology,* 4, 285.

Cash, M. (2013): "Cognition without borders: 'Third wave' socially distributed cognition and relational autonomy". *Cognitive Systems Research,* 25–26, 61–71.

Clark, A. (2008): *Supersizing the Mind: Embodiment, Action and Cognitive Extension.* Oxford, UK: Oxford University Press.

Clark, A. (2009): "Spreading the joy? Why the machinery of consciousness is (probably) still in the head". *Mind,* 118(472), 963–993.

Clark, A. (2010): "Coupling, constitution and the cognitive kind: A reply to Adams and Aizawa". In Menary, R. (ed.), *The Extended Mind.* Cambridge, MA: MIT Press, 81–99.

Clark, A. & Chalmers, D. (1998): "The Extended Mind". *Analysis* 58: 1, 7–19.

Colombetti, G. & Krueger, J. (2015): "Scaffoldings of the affective mind". *Philosophical Psychology.* https://doi.org/10.1080/09515089.2014.976334

Couch, M. B. (2011): "Mechanisms and Constitutive Relevance". *Synthese* 183 (3): 375–88.

Craver, C. F. (2007a): *Explaining the Brain: Mechanisms and the Mosaic Unity of Neuroscience.* Oxford: Oxford University Press.

Craver, C. F. (2007b): "Constitutive explanatory relevance". *Journal of Philosophical Research* 32: 3–20.

Craver, C. F. and Bechtel, W. (2007): "Top-down causation without top-down causes". *Biology and Philosophy* 22: 547–63.

Crick, F. & Koch C. (2003): "A Framework for Consciousness". *Nature Neuroscience,* 6(2): 119–26.

Dainton, B., (2008): "The Experience of Time and Change". *Philosophy Compass* 3/4, 619–38.

Dainton, B. (2018): "Temporal Consciousness". *The Stanford Encyclopedia of Philosophy.* Edward N. Zalta (ed.) <https://plato.stanford.edu/archives/win2018/entries/consciousness-temporal/>.

Fisk, J., Lackner JR., & Dizio, P. (1993): "Gravitoinertial force level influences arm movement control". *J Neurophysiol* 69 (2): 504–511.
Foster, J. (1991): *The immaterial self: A defence of the Cartesian dualist conception of the mind*. London: Routledge.
Gallagher, S. (2013): "The socially extended mind". *Cognitive Systems Research*.
Gallagher, S. (2018): "New Mechanisms and the Enactivist Concept of Constitution". In M. P. Guta (ed.): *The Metaphysics of Consciousness*, 207–220. London: Routledge.
Gillett, C. (2013): "Constitution, and multiple constitution, in the sciences: Using the neuron to construct a starting framework". *Minds & Machines* 23: 309–337.
Goldin-Meadow, S., Nusbaum, H., Kelly, S. D, & Wagner S. (2001): "Explaining math: Gesturing lightens the load". *Psychological Science* 12(6): 516–22.
Harbecke, J. (2010): "Mechanistic Constitution in Neurobiological Explanations". *International Studies in the Philosophy of Science*, 24(3): 267–285.
Harbecke, J. (2015): "Regularity Constitution and the Location of Mechanistic Levels". *Foundations of Science* 20(3): 323–338.
Hurley, S. (1998): *Consciousness in Action*. Cambridge, MA: Harvard University Press.
Hurley, S. (2010): "The Varieties of Externalism". In Menary, R. (ed.), *The Extended Mind*. Cambridge, MA: MIT Press, 101–153.
Husserl, E. (1991): *On the Phenomenology of the Consciousness of Internal Time*. (1893–1917). Edited and translated by Brough, J. B. Dordrecht: Kluwer.
Kaiser, M. (2018, in press): "Individuating Part-Whole Relations in the Biological World". In O. Bueno, R.L. Chen, and M.B. Fagan (eds.), Individuation, Process and Scientific Practices. Oxford University Press.
Kaiser, M. & Krickel B. (2017): "The Metaphysics of Constitutive Mechanistic Phenomena". *Brit. J. Phil. Sci.* 68: 745–779.
Kaplan, D. M. (2012): "How to Demarcate the Boundaries of Cognition." *Biology and Philosophy* 27: 545–70.
Kirchhoff, M. D. (2013): "Extended cognition and the causal-constitutive fallacy: In search for a diachronic and dynamical conception of constitution". *Philosophy and Phenomenological Research* 90 (2): 320–60.
Kirchhoff, M. D. (2015): "Cognitive assembly: Towards a diachronic conception of composition". *Phenomenology and the Cognitive Sciences*, 14(1), 33–53.
Kirchhoff, M. D. (2016): "From mutual manipulation to cognitive extension: Challenges and implications". *Phenomenology and the Cognitive Sciences*, 16, 863–878.
Kirchhoff, M. D. & Kiverstein, J. (2019): *Extended Consciousness and Predictive Processing: A Third Wave View*. London, UK: Routledge.
Kiverstein, J. & Arstila, V. (2013): "Time in Mind". In Dyke, H. & Bardon, A. (eds.) *A Companion to the Philosophy of Time*. Wiley 444–469.

Krickel, B. (2018): "Saving the Mutual Manipulability Account of Constitutive Relevance". *Studies in History and Philosophy of Science*. Part A 68: 58–67.
Krueger, J. (2018): "Schizophrenia and the Scaffolded Self". *Topoi*. https://doi.org/10.1007/s11245-018-9547-3
Levy, N. (2011): "Neuroethics and the extended mind". In J. Iles & B. Sahakian (Eds.), *Oxford Handbook of Neuroethics*, Oxford: Oxford University Press, 285–294.
Machamer, P., Darden L., & Craver C. (2000): "Thinking about Mechanisms". *Philosophy of Science* 67, 1–25.
Menary, R. (2007): *Cognitive Integration: Mind and Cognition Unbounded*. New York: Palgrave Macmillan.
Menary, R. (ed.) (2010a): *The Extended Mind*. Cambridge, MA: MIT Press.
Menary, R. (2010b): "The holy grail of cognitivism: A response to Adams and Aizawa". *Phenomenology and the Cognitive Sciences*, 9, 605–618.
Noë, A. (2004): *Action in Perception*. Cambridge, MA: MIT Press.
Noë, A. (2006a): "Experience without the head". In Gendler, T. S. and Hawthorne, J. (eds.) *Perceptual Experience*. New York: Oxford University Press. 411–434.
Noë, A. (2006b): "Experience of the world in time". *Analysis*, 66(289), 26–32.
O'Regan, K. & Noë, A. (2001): "A sensorimotor account of vision and visual consciousness". *Behavioral and Brain Sciences*. 24(5): 939–973.
O'Shaughnessy, B. (2000): *Consciousness and the World*. New York: Oxford University Press.
Palermos, S. (2014): "Loops, constitution, and cognitive extension". *Cognitive Systems Research*, 27, 25–41.
Pepper, K. (2014): "Do sensorimotor dynamics extend the conscious mind?". *Adaptive Behavior*, 22(2), 99–108.
Rescher, N. (2000): *Process Philosophy: A Survey of Basic Issues*. Pittsburgh: University of Pittsburgh Press.
Roberts, T. (2012): "Taking responsibility for cognitive extension". *Philosophical Psychology*, 25, 491–501.
Ross, D., & Ladyman J. (2010): "The Alleged Coupling-Constitution Fallacy and the Mature Sciences". In Menary, R. (ed.) *The Extended Mind*. Cambridge, MA: MIT Press, 155–166.
Seibt, J. (2009): "Forms of emergent interaction in General Process Theory." *Synthese*, 166, 479–512.
Silverman, D. (2013): "Sensorimotor enactivism and temporal experience". *Adaptive Behavior* 21(3), 151–158.
Sprevak, M. (2009): "Extended Cognition and Functionalism". *Journal of Philosophy*. Vol 106, no. 9, 503–527.
Sprevak, M. (2010): "Inference to the hypothesis of extended cognition". *Studies in History and Philosophy of Science*, Part A, 41, 353–362.

van Gelder, T. J. (1998): "The dynamical hypothesis in cognitive science". *Behavioral and Brain Sciences* 21(5): 615–28.
van Gelder, T. J. (1995): "What might cognition be, if not computation?" *Journal of Philosophy*, 92, 345–381.
Varela, F. J. (1999): "The specious present: A neurophenomenology of time consciousness". In J. Petitot, F. J. Varela, B. Pachoud, and J-M. Roy (eds.), *Naturalizing Phenomenology: Issues in Contemporary Phenomenology and Cognitive Science*. Stanford: Stanford University Press, 266–314.
Virgo, N., Egbert, M. & Froese, T. (2011): "The Role of the Spatial Boundary in Autopoiesis". In: G. Kampis, I. Karsai & E. Szathmary (eds.). *Advances in Artificial Life: Darwin Meets von Neumann*. Berlin: Springer. 240–247.
Wasserman, R. (2004): "The Constitution Question". *Noûs* 38: 693–710.
Weiskopf, D. A. (2010): "The Goldilocks problem and extended cognition". *Cognitive Systems Research*, 11, 313–323.
Wiggins, D. (1968): "On Being in the Same Place at the Same Time," *Philosophical Review*, 77(1): 90–95.
Woodward, J. (2003): *Making Things Happen: A Theory of Causal Explanation*. New York: Oxford University Press.
Ylikoski, P. (2013): "Causal and Constitutive Explanation Compared". *Erkenntnis* 78(2): 277–297.

CHAPTER 4

Tools as Extenders: The Pathway to Functional Incorporation

4.1 Introduction

In this chapter, I will outline the preconditions for my position concerning extended conscious mind (ECM). I will state what is needed for there to be a case of ECM, namely which requirements need to be fulfilled. This chapter is more about setting the limits of what is constitutive specifically in ECM and is more pragmatically and empirically oriented than the previous chapter. However, the underlying question remains the same: how to distinguish constitutive components from etiological or causal background conditions. In this chapter, the question concerns my own position: I suggest how the lines should be drawn with ECM. The borderlines I draw are much narrower than in many other, more general 4E-accounts. Whereas the second chapter provided arguments why we should accept ECM and the third chapter surveyed how the relation can be constitutive, the present chapter is more descriptive. It doesn't aim to *argue* for ECM or for its constitutive character but rather show what is meant by (constitutive) extension in ECM. The criteria I will set for ECM should be taken as *epistemic tools* rather than ontological statements—the criteria itself are not answers to the question of what constitutes the phenomena they are about but rather tools that help us to see what belongs to those constituents.

P. Telakivi, *Extending the Extended Mind*, New Directions in Philosophy and Cognitive Science,
https://doi.org/10.1007/978-3-031-35624-7_4

The chapter will proceed as follows. First, I will demarcate different degrees of extensions, from mere non-functional momentary extension to functional, sensorimotorically controlled incorporation. This demarcation, even though not categorical, but rather spectrum or continuum like, is needed for the later parts of the chapter. After presenting the spectrum of extensions, I will review the literature on what *glue & trust* criteria have been proposed for different sorts of externalisms so far, and subsequently give my own glue & trust conditions for ECM. Finally, I will discuss one concrete example that I suggest fulfils my glue & trust conditions, namely *sensory substitution* technology.

4.2 The Spectrum of Extensions

I will differentiate between four "types" of extension—from weaker to stronger, from fleeting to more robust. This taxonomy should not be read as a categorisation of different kinds or classes of extensions but rather as a scale or spectrum where the borders between different groups are not steadfast. In the literature, the differences in extension have usually been presented in pairs of two, for example *extension* vs. *incorporation* (De Preester and Tsakiris 2009; Thompson and Stapleton 2009; De Preester 2011) and *use* vs. *incorporation* (Clark 2008). The distinctions have been justified with the help of neurophysiological research on how tool use affects body schema (Maravita and Iriki 2004; Berti and Frassinetti 2000). According to Thompson and Stapleton (2009), the difference between extension and incorporation is that with the former, we extend our abilities, and with the latter, we not only extend our abilities but also a phenomenologically different status is gained.[1] This differentiation is closest to the distinction I draw between the first two types (momentary extension and integrated extension) together contrasted with the last one (functional incorporation).

Another distinction made in the literature is between *dispositional* and *occurrent extensions* (see the discussion in the second chapter, and Colombetti and Roberts 2014). The former refers to standing non-occurrent mental states (such as a standing belief of a museum's address written down in one's notebook) and the latter refers to cases of

[1] They mention that their distinction is indebted to De Preester's work. De Preester and Tsakiris (2009) and De Preester (2011) have probed deeper into the topic and revealed the complexity of the matter.

temporary states that occur at a certain moment (such as the act of counting realised with pen and paper). As my aim is to examine experience, and because experience is always occurrent, all my examples here fall under occurrent extension. Thus, dispositional extension is not discussed in this chapter.

Distinctions have also been made within the notion of incorporation. Colombetti (2016) notes that Merleau-Ponty deals with two kinds of incorporation. She calls them *habit incorporation* and *object incorporation*. Merleau-Ponty's (1945, 166) example of habit incorporation is an experienced typist who doesn't have to pay attention to the keyboard. Her fingers 'know' what to do. The bodily awareness (gained through habit) of where each letter is on the keyboard is similar to our habitual awareness of where our limbs are. Object incorporation, on the other hand, refers to cases where *material objects* are incorporated into the body schema, for example, a blind person's cane. However, my differentiation within the notion of incorporation is different from Merleau-Ponty's (and Colombetti's). Nonetheless, it is good to recognise the existence of habit incorporation as well and the different ways the term incorporation can be used.

I think it is useful to have an even more fine-grained taxonomy than any of these, which is why I will present four degrees or dimensions of extensions. I have labelled them as follows: (1) *momentary extension*, (2) *integrated extension* (or sometimes just *integration* for short), (3) *prosthetic incorporation*, and (4) *functional incorporation*. The taxonomy will reveal the differences in the degree of attachment: how firmly the external part is part of the system (this doesn't mean the level of physical attachment). It is important to notice that all four groupings fall under (constitutive) extension.[2] They are different ways for an external element to be a constitutive component of a mental state, process, or system. Hence, the distinction between extended and embedded is outside of this setting—all four dimensions belong to the class of *extended* (as differentiated in the first chapter, Sect. 1.3).

The grading of extensions also helps us to see the difference between EM and ECM. Revealing the difference is not straightforward because all types of extensions describe occurrent states that are both cognitive and

[2] Even in prosthetic incorporation, we can say that the external component is constitutive for the certain bodily function it serves—that function just doesn't fall into EM or ECM because the function in question is neither cognitive nor experiential.

experiential, and as such all of them (except prosthetic incorporation) could be considered to belong to both EM and ECM. This again shows how it is difficult to separate occurrent EM from ECM, as was argued in Chap. 2.

However, I classify EM as something that belongs to the group of *integration*. Even though the type of integration includes experiential qualities, it is not enough to qualify as ECM as I define it. On the other hand, even though functional incorporation surely contains cognitive processing as well, it only qualifies as a group of ECM and not of EM because the experiential aspects are what differentiates it.

The main demarcating question is rather *what* is extended: an experiential state or process contrasted with an experiencing subject. The latter is needed for there to be a genuine case of ECM. This differentiation is similar to Clark and Chalmers's (1998) when they separate *cognitive processing* from the *mind* (by *mind* they mainly refer to *beliefs*). The former involves individual acts such as using a rotation button while playing Tetris, and the latter involves the whole of Otto's memory/belief system. In a similar vein, in my differentiation, the first two degrees describe the extension of experiential (and cognitive) states or processes (such as the act of counting using pen and paper), and the fourth degree describes the extension of a larger system (such as a person using a blind person's cane). The third group, prosthetic incorporation, is included only for clarificatory purposes; it is an instance neither of EM nor of ECM. Nevertheless, it is necessary to distinguish this type of extension from the fourth one. As said, these levels of extension should be taken as epistemic tools rather than ontological statements. Their purpose is to explain the differences in how an external object may become part of a coupled system that realises an experiential state.

An important aspect in the demarcation of these levels of extension is the notion of *functionality*. It does not refer to *functionalism* as it is understood in analytic philosophy of mind, where it is a close ally with cognitivism. Obviously, I do not oppose functionalism's core thesis when it is understood in a loose sense that stresses the multiple realisability of mental states—providing that the multiple realisability can instantiate in brain, body, and environmental elements.[3] This, however, is not what I am aim-

[3] This much is obviously true for anyone who supports any kind of externalist view about the mind. Still, functionalism (e.g., Wheeler's psychofunctionalism) as a doctrine in analytic philosophy of mind is problematic because it promotes a sharp separation between cognition and consciousness and hence should not be confused with the views presented here.

ing at with the notion. Instead, I refer to something that is taken as part of the *functional system*—not only that it has a function that it fulfils. This could be said to be an enactivist reading of function(alism). Being a physical part is not enough; the tool has to be under the sensorimotor control of the subject, and it has to become part of the transparent subjective point of view. It goes without saying that my notion of functional incorporation is very much influenced by Merleau-Ponty's (1945) views on incorporation.

Another aspect that needs some further clarification is the concept of *tool*. It is perhaps the most central concept in this chapter—extension occurs in and by means of tools. For the position I am arguing for, the tool needs to be a material object. Thus, habits, cultural practices, and so on are ruled out from my use of the concept. However, I, by no means, oppose the idea that habits and cultural practices can also be ways of extension (e.g. Merleau-Ponty's habit incorporation). They are just not what I am after with my notion of ECM, which is a narrower concept that includes only incorporation, that is, the experiencing body extended with physical appliances. This is related to a quite substantial difference between my account and Kirchhoff and Kiverstein's (2019, 2020) account of ECM. In their account, also cultural environments can be part of the constitutive basis of conscious experiences: the way we are *phenomenally attuned* to the world is constrained by cultural habits and practices. I don't disagree with their position, but I would rather place this phenomenon under another E: embeddedness.

So, in this book, tools as extenders of conscious experiences are material tools that we can hold or in other ways physically control. Frédérique De Vignemont (2018) refers to Beck's (1980) landmark book on animal tool use and spells out that tools extend our *motor*, *sensory*, or *spatial* capacities. Helena De Preester's (2011, 121) list is slightly different—according to her, tools extend *motor*, *sensory*, or *cognitive* capacities.[4] A *tool* does not refer to whatever object but only to "unattached external objects that one actively manipulates—and not simply holds—for functional purpose" (de Vignemont 2018, 2).

De Vignemont's and De Preester's definitions leave some unanswered questions. For example, how *unattached* do the objects need to be? Even

[4] I think it is problematic to distinguish 'cognitive' from the other capacities. All the instances of tool use that I use as examples here are not purely cognitive but are also sensory or motor as well.

though hearing aids and sensory substitution devices are rather firmly attached to the body, I think they should be treated as tools. I suggest that the level of attachment is not a crucial question when we need to decide whether something is or is not a tool. Neither is the fact whether the tool is replacing (e.g. an artificial limb), substituting (e.g. a sensory substitution device), enhancing (e.g. a hearing aid), or creating (e.g. a hammer or a robotic sixth finger)[5] a body part or function. However, the location of the putative tool—whether it is located inside or outside the skin—might affect the criteria because the location is related to the possibility of the subject's motor control.[6] Moreover, if taken literally, de Vignemont's definition would also lead to a situation where if somebody was just holding, say, a fork or a hammer, they wouldn't count as tools.

However, I agree that active manipulation, adaptation, and a functional goal-drivenness are necessary for what we are looking for. Merely holding an object (even if it was a hammer) doesn't count as *tool use* in the sense we are after here. To understand what is meant by 'tool', we need to look at its different roles for the user. These issues show that we cannot actually define the meaning of tool without its *use*. Whether something counts as a tool depends on how it is used and on its relation to the tool user. The level of attachment and entrenchment varies, and simply talking of 'tool use' still leaves open what kind of relation is in question. The following sections aim to shed light on the different dimensions of tool use as instances of extension, integration, and incorporation.

4.2.1 Momentary Extension

Consider an everyday task carried out with a tool. For example, imagine me cutting someone's hair with scissors. I am not an expert, and the result would not look pretty, but I am still capable of shortening hair because I can use scissors. The relation between the scissors and me is only temporary, but it is goal driven, and we can say that a sensorimotor *state* is extended. We are surrounded by tools that are potential for short-term, momentary extension—pens, spoons, hammers, axes, rakes, spades, tennis rackets, and so on. They are the very common "artifacts that we use and

[5] For the robotic sixth finger, see Meraz et al. (2018).

[6] I am not saying that a device located under the skin (endoprosthesis) could not count as a tool *in principle* or that the skin would be an absolute demarcating barrier. I just want to draw attention to practical limitations. What matters is the possibility of taking the tool under one's control, and it is more difficult with an endoprosthesis. I will discuss this in the section on *prosthetic incorporation*.

control in order to extend our abilities" (Thompson and Stapleton 2009).[7] As Clark (e.g. 2008) has clearly shown, we off-load cognitive burden to the environment (to tools and by means of tools) and create new ways of cognising with tools. However, this off-loading does not only happen in the cognitive sphere; it is also very much a sensorimotor and perceptual phenomenon. It extends the sensorimotor body.

This type covers incompetent tool users, that is, people who are beginners in using some tool. For example, think about somebody starting to practise to play the violin. The instrument fulfils its role, but the violin doesn't become part of the (transparent) system: it is not rooted to the system and made one's "own". Of course, sometimes also the tools we have used for a long time can stay on mere momentary extension level; for instance, consider somebody who has played the violin for ten years, but still feels quite uncoordinated and detached from her violin, unable to express what she would like to express with it. A fluent tool user is a different story and belongs to the type of integration or incorporation—depending on which aspects we are looking at. As I have stressed, the borders of these degrees of extension are not set in stone, and each case might require individual conditions.

Technically speaking, perhaps we should demarcate a group of extension even below this one: *mere tool possession* without extension. Consider somebody just holding a tool without intentionally manipulating anything with it. The task might still get accomplished, but there is no sense of agency or control over the tool included. For example, consider a small child holding a knife—she might accidentally cut something without realising and without aiming to cut anything. Or somebody might move a pen on paper while doing something else that captures her awareness completely. However, as the reason for presenting these four types is that they are different kinds of instances of extension, *mere tool possession without extension* wouldn't belong to that setting.

Momentary extension is a predecessor and a more fleeting version of integration. In momentary extension, the tool extends only certain *states* (e.g. cognitive, sensorimotor, or perceptual), but the whole *system* (of which the aspect can be, e.g. cognitive, sensorimotor, or perceptual) is left untouched. Integration, on the other hand, extends the whole system. So the difference between these two types is about what aspect or entity we are asking about. Moreover, the type of momentary extension is not

[7] I would not disagree that some techniques in general—such as writing—also count as instances of extension, but I limit the discussion to concrete material objects.

functional in that specific way in which I use the notion (which will be discussed further later). *Momentarily extended* tools surely have a function in a general sense of the term (e.g. the scissors succeed in their function of cutting hair). But this is not enough to count as 'functional' in the way I am using the concept.

Even though I think that even with a short-term tool use externalisation may occur, I admit that the fleeting nature of tools that are used only rarely or sometimes might be problematic for the theory. Already in their seminal paper, Clark and Chalmers (1998) restricted their theory: they argued that the notebook must play a *constant* and *stable* role in Otto's life. I am going to restrict my position in a similar way. Thus, the type of momentary extension is not enough for something to count as ECM.

4.2.2 Integrated Extension

This time, think about Sweeney Todd, who cuts his customers' hair (and throats) with an expertise he had gained during his years as a barber in Fleet Street. The razor-sharp scissors move effortlessly, without the need to pay attention to where his hand ends and the tool begins. Whereas an unexperienced scissor user won't notice the difference between one pair of scissors and another, Sweeney Todd has created a special relationship with his scissors that can even be reciprocal. His way of using scissors has partly created some of their properties, and on the other hand, the special features of the scissors have shaped how the barber cuts with them.

This kind of coupling is more transparent, stable, and individualised than momentary extension. Sweeney Todd doesn't have to concentrate on the technical details of the cutting act or the scissors, but he has a direct, transparent access to his target in the world (hair) and how he wants it to be cut and shaped. In a similar way as a skilful artist paints a portrait without needing to look at her model before every brush stroke (see Noë 2015), a skilful barber works with hair. Following Noë's idea, this action is an interactive, dynamic process that more closely resembles dancing than it does the representation of ideas. I suggest that the scissors are *integrated* into Sweeney Todd's sensorimotor, perceptual, and cognitive systems.[8] In his case, not only a temporary occurrent sensorimotor *state*

[8]With *integration*, it is not evident *into what* the external tool is actually integrated—it could be the mind, the cognitive system, an organism, the self, and so on. Instead, in the case of in*corporation*, the target is quite self-evidently the *corpus* (the body) and the bodily stance or point of view.

extends, but he and his scissors together form a new whole—a coupled system whose function is to cut hair (and throats).

This type of extension is about the same phenomenon that was described in the previous section, but it is more permanent and affects a larger area than momentary extension. Otto could be seen as an example of this kind: an external element (the notebook) is *integrated* into the system (Otto's memory capacity). It is not only about a specific museum visit, but about his whole memory system. Also the way most of us use smartphones these days counts as an example of integration. The phone has been taken into our system: it is trusted, it has gained an almost transparent status, and if it is taken away, many of our cognitive functions suffer. As a guide in differentiation, we can use similar criteria with Clark and Chalmers (1998). Especially their first glue & trust criterion—constancy, durability, and reliability—is what differentiates integration from momentary extension. The difference between momentary extension and integrated extension is a question of degree and, as said, should be considered individually in each case.

The term 'integration' or 'integrated' has been used in the literature, for example, by Menary (2007, 2010), who defines EM in terms of integration. He uses the term for differentiating his view from arguments based on cognitive *parity*. According to him, since EM does not rely on the parity principle, the external elements do not have to be similar or have similar functions with the biological ones—they do not have to be "functionally equivalent". Instead of parity, the external elements are integrated into the system in a complementary way. Menary (2010, 235) writes that "it is because the external vehicles provide a different kind of functionality and because they can coordinate with internal processes that they are integral parts of our cognitive systems".

However, I use the term in a slightly different way. I accept Menary's view and agree that his account of integration is a better way to understand externalist views than the first-wave views based on parity principle. However, this is not what I am after with the notion of *integration*. My aim is to differentiate integration, on the one hand, from momentary extension, where the external element doesn't get integrated into the system, and on the other, from incorporation, where bodily ownership is required. Integration, according to my reading, is hence a stronger relation than momentary extension and a weaker relation than incorporation.

Integration is a type of extension that happens in EM. Even though an experiential side is included, some aspects are not present that are needed for ECM. For example, presumably the body schema doesn't get extended in this type of extension. This might give empirical methods to track the difference between integration and incorporation (for neurophysiological studies on the extension of body schema, see Maravita and Iriki 2004; Berti and Frassinetti 2000). I will list the other differentiating aspects in the section on functional incorporation. However, before that, let us look at another type of incorporation.

4.2.3 Prosthetic Incorporation

Consider a patient using a cardiac pacemaker to maintain a healthy heart rate. It is implanted in her body and will become a non-biological part of the body. Or consider another endoprosthesis, that is, a prosthesis that is located inside the body, for example, in the hip where a dysfunctional joint is replaced with an artificial one. The pacemaker and the prosthetic hip enable bodily functions that keep the patient alive or at least improve the quality of her life significantly. We can also think of a transplant (e.g. heart or kidney)—a biological version of prosthetic incorporation.

Prosthetic incorporation is *mechanical* rather than integrated or functional in the sense I am using the term. An endoprosthesis is functional only in the narrow sense that it becomes part of the physical, bodily functions. It is not functionality as I define it in this taxonomy of extensions. The crucial difference between a prosthesis that is located inside the body and a tool that the subject is using is that the former is not under the control of the experiencing subject, whereas the latter is. Moreover, we cannot feel the world through an endoprosthesis; the transitive nature of a tool is lacking. Thus, an endoprosthesis or other prosthetically incorporated object is not a tool.

A terminological clarification is needed here. I use the term 'prosthesis' to refer only to cases that fall under prosthetic incorporation. I don't call them 'tools' since they don't satisfy the requirements of how I define 'tool': they are not unattached objects taken under the ownership, control, and point of view of the user. These concepts are not used consistently in the literature; for example, 'prosthesis' sometimes refers to any device that the subject can *use* or *incorporate*, and hence also includes what I refer to as tools (e.g. De Preester 2011).[9]

There are two different ways for an object to belong to the group of prosthetic incorporation. First, when they are *replacing* body parts or functions, as was the case with the pacemaker and artificial hip (especially internal body parts, i.e. endoprostheses, but the distinction between external and internal replacements is not rigid but may vary depending on the case and what aspect of it we are interested in).[10] Second, the object can create an extra part or function that does not replace any pre-existing function. This is possible at least in principle and in the case of Australian performance artist Stelarc and his many prosthetic objects, such as a statue in his stomach.[11]

Even though the type of prosthetic incorporation is non-functional, it still deserves to be called incorporation, instead of momentary extension or integration, because an external, non-biological element is taken as a part of the (physical) body. Furthermore, this kind of extension should still technically go under the label of *constitutive* because artificial body parts clearly work as constitutive components for the *bodily* functions in question. However, prosthetic incorporation is not very interesting for my purposes—it is not the kind of extension that is needed for ECM. Still, this type's existence is important to acknowledge because it shows that incorporation can be divided into two: prosthetic and functional incorporation, where the former is mechanical and passive, and the latter active and goal driven. Apart from showing what kind of incorporation is needed for ECM (i.e. functional rather than prosthetic), pinpointing the group of prosthetic incorporation also sorts out one potential misunderstanding that could be used as a critical note against ECM, namely the claim that ECM's proponents allegedly argue for ECM on the grounds of prosthetic

[9] I will discuss artificial limbs briefly in the section about functional incorporation. It is not a simple question whether they should be treated as prostheses or tools.

[10] This group also contains thought experiments where neurons are replaced by silicon circuits. Apart from philosophers' armchairs and science fiction, this kind of experiment has in fact been conducted with California spiny lobster (even though with only one of its neurons). I will discuss lobsters and silicon circuits soon.

[11] Yet another interesting question is chemical altering of the brain (e.g. with drugs or antidepressants) and whether that would belong to the group of prosthetic incorporation or not. However, the dosage can at least, in principle, be under control of the user (but in practice, control might be diminished).

incorporation.[12] One passage by Andy Clark (2009) initiated the discussion, and Katalin Farkas (2012) and Karina Vold (2015) continued.

> [I]magine a case in which a person (call her Diva) suffers minor brain damage and loses the ability to perform a simple task of arithmetic division using only her neural resources. An external silicon circuit is added that restores the previous functionality. Diva can now divide just as before, only some small part of the work is distributed across the brain and the silicon circuit: a genuinely mental process (division) is supported by a hybrid biotechnological system. That alone, if you accept it, establishes the key principle of [EM]. (Clark 2009)[13]

Farkas (2012)—quite justifiably—writes that if one accepts this much, there is no reason to leave conscious experience out. (I agree with her about this, but I don't accept Clark's original claim that this would be enough to establish EM.) In order to show that Clark's method is not restricted to the cognitive sphere, but that it encompasses consciousness as well, Farkas modifies Clark's example slightly:

> [C]onsider the example of Hera, who suffers minor brain damage and loses the ability to process auditory signals using only her neural resources. Therefore she does not have auditory experiences anymore. An external silicon circuit is added that restores the previous functionality. Hera's experience of hearing is completely restored: but now a genuinely conscious event (an auditory experience) is supported by a hybrid biotechnological system. (Farkas 2012, 443–444)

Farkas's main reason for using the *silicon chip* argument is to show Clark's inconsistency and not to argue directly for ECM with it. Vold (2015), on the other hand, directly uses the silicon chip argument for ECM. She asks us to imagine a situation where a patient's brain functions are deteriorating and doctors start replacing her biological neurons with artificial ones that are programmed to behave identically with the original ones:

[12] 'Prosthetic incorporation' is my term, though. None of the writers referred to here use the term.

[13] In Clark's defence, he wrote this passage as an answer to Fodor's critique to his book *Supersizing the Mind*, and the example of Diva is tailored for that purpose.

> Now imagine that instead of entering your skull to replace your deteriorated neurons where they are, doctors consider it preferable and less intrusive to the biological shell to do as much as they can externally. So doctors tell you it will be safer to attach your remaining well-functioning neurons through tiny electrical nodes to an implant that threads a wire out your ear. They then attach this wire to an external device that contains your programmed silicon chips and which attaches to your person. We can imagine that after the operation you come to with a small wire now reaching out of one of your earholes. Attached to this wire is a small device that hugs your ear, much like a hearing aid. Doctors call it an iCog. (Vold 2015, 27)

By placing the silicon chips outside the body, Vold probably wanted to stress the externalist side—which is after all the ultimate goal behind all of these examples. However, the thought experiment doesn't get any more convincing by relocating the vehicle outside the skin and skull. It is irrelevant where the artificial neurons are *located* or *what they are made of*, as the proponents of the 4E-framework have promoted. Instead the important question is can we actively interact with them—use them, do things with them—and as a result take them into our system as genuine parts of ourselves. Externalism is not established by stipulating a logical possibility of replacing neurons with silicon chips and placing them outside the head. I claim that these thought experiments do not promote externalism at all because they have misunderstood the most crucial features of EM and ECM.

Thus, the silicon chip argument is inadequate and misguided for two reasons. First, because it is a speculative thought experiment.[14] The Otto thought experiment was at least based on existing, everyday "technology". However, an even more serious problem is the second one: the silicon chip argument belongs to the group of *prosthetic incorporation*. I will explain in the following why prosthetic incorporation is incapable of backing up EM or ECM.

[14] Even then, something like this has been done in empirical research. A group of scientists have successfully managed to replace one neuron of a California spiny lobster with an artificial, electronic one that functioned well with the biological ones, and they aim to gradually replace more neurons (see Szucs et al. 2000, and discussion in Clark 2008). There are undoubtedly more examples, and we will certainly see more coming with the technology developing.

David Chalmers (2019)[15] ends up reaching a similar conclusion, even though I find his path of argumentation wanting, because he doesn't differentiate between prosthetic and functional incorporation (actually he doesn't talk about any 'incorporation' but instead only about 'extension' without contrastive subtleties). Nevertheless, he also concludes that the silicon chip argument is weak. As a clear indication of the weakness, he mentions that even the best-known opponents of the externalist views accept the silicon chip argument: Adams and Aizawa (2008) and Rupert (2009) have all declared that they can agree with science-fiction-type thought experiments that include imaginary future technology, such as silicon chips replacing brain functions. I agree with Chalmers that already this alone shows that there is something wrong with the silicon chip arguments and the like. Somewhat surprisingly, he finds a solution that follows a well-trodden path. Chalmers's solution to the inadequacy of the silicon chip arguments is virtually *sensorimotor enactivism*, although he doesn't call it so. He states that what is distinctive for EM is "interacting via perception and action", and that is what makes Otto and his notebook as well as rotation in Tetris count as EM, whereas the silicon chip examples are ruled out. He gives an updated definition of EM:

> A subject's cognitive processes and mental states can be partly constituted by entities that are external to the subject, in virtue of the subject's sensorimotor interaction with these entities. (Chalmers 2019)

It is easy to agree with Chalmers's statement—it is the core assertion that I have made throughout this work: extension is based on an interaction relationship via perception and action between the subject and the external element. Without that relationship, there is no extension in the sense that it is understood in EM or ECM.

Let us recapitulate why prosthetic incorporation is insufficient for ECM. To start with, the prosthetically incorporated objects are not 'tools', as I have described a 'tool' based on De Preester's (2011) and De Vignemont's (2018) definitions: they are not "unattached external objects that one actively manipulates" to extend our "motor, sensory or spatial capacities". Further, there is no sensorimotor control over the

[15] As I mentioned in the first chapter, Chalmers has mostly stayed out from the extended mind debates since their original article with Clark. This 2017 paper, however, is the most notable exception.

prosthetically incorporated objects—the subject is not in charge—and hence a sense of agency over the "tool" will not develop. From these reasons it follows that the subject cannot manipulate her surroundings with the device; it is beyond her control.

Unlike the previously presented ways of extensions (and functional incorporation, which will be presented next), prosthetic incorporation is not on the same scale: a cardiac pacemaker either works or doesn't work; the patient doesn't have active control over it (a surgeon is needed to replace it). The three other ways of extension require getting used to it by practice, *attuning* to it: a process in which the tool is taken as one's "own". A successful coupling and extension only happens as a result of this effort. The level of coupling between a violinist and her violin is always a matter of degree. Prosthetic incorporation is different in this sense, and its functioning is quite out of reach of the subject's training.

Thus, prosthetic incorporation is not only "not enough" as an argument for ECM, but it is an altogether wrong direction to look for it. I do not think prosthetic incorporation provides any support *at all* for ECM. All the defining features of ECM are missing. Even if we could replace all of the neurons and brain functions with silicon chips (or some other nonbiological material), that would not yet tell us anything about ECM being true or not.

Furthermore, apart from prosthetic incorporation, we could also distinguish a group of *prosthetic extensions* that occur on the surface of the body, rather than under it, and mainly affect the body image. This group includes wearables such as clothes, headphones, scarfs, and other headgear in different religions, facemasks, jewellery, watches, and so on. Or—getting back to hair-related examples—consider somebody using a wig because she has lost her hair. Material or sensorimotor function might not be very significant, but social function might be crucial. These kinds of prosthetic extensions take part in building our identities and social selves.[16] Even though the subject has control over the prosthetically extended elements, this is not what I am after with ECM, because they don't extend the body schema. Let us now turn to the positive story. What features are thus required for ECM, if not prosthetic incorporation or extension?

[16] I want to thank Chiara Cappelletto for pointing out that this group is separate from my other levels of extension.

4.2.4 *Functional Incorporation*

Finally, consider Edward Scissorhands. For him, scissors are part of his body, and he experiences the world through them in a way that is not possible for even the most competent of Fleet Street barbers. Is there a difference in kinds of extensions between the cases where the tool is used fluently (Sweeney Todd) and when the tool is part of the body (Edward Scissorhands)?[17] Of course, the latter case is science fiction (and the former is fiction, but that is not relevant). Even though scissors as body parts exist only in science fiction, this comparison still reveals a crucial question: how firmly and in what sense does a tool need to be attached to the body for it to be considered *functionally incorporated*?[18]

From everything we have learned so far, it has become clear that the tool doesn't have to be biological. Nor does it have to be attached permanently (or not even firmly in a physical sense) to the body, as Edward's scissor hands are or as an artificial cardiac pacemaker is. To understand what is meant by this fourth group of extension, we need to look at both terms—*functional* and *incorporated*. In the following, I will discuss these two concepts—separately and together—and aim to show how functional incorporation differs from the other three types of extensions.[19]

Let us first look at the term *incorporated*. Here, it is understood differently than in the prosthetic version of incorporation—it doesn't need to involve taking the external tool physically inside the boundaries of the body (as in the previous incorporation). Rather, the external part needs to be taken into the *bodily point of view*—so that the subject doesn't experience the new external part as an object, but that she experiences the world through it. The notion of *body schema* is a central feature when defining functional incorporation. It is differentiated from *body image*, and this

[17] The idea for contrasting Sweeney Todd with Edward Scissorhands is borrowed from Thompson and Stapleton (2009). However, they don't develop the example further (they don't talk about unskilled scissor users or wigs); it is just briefly mentioned to stress the importance of intimate coupling and bodily regulation for incorporation relations. Their paper is to thank for the inspiration behind my hair-related examples, though.

[18] The Australian performance artist Stelarc could almost be called a real-life example of Edward Scissorhands. He is famous for enhancing his body by technical means, for example, he has transplanted an ear in his hand (that is "partly surgically constructed, partly cell-grown") and a robotic third arm (See Atzori and Woolford 1995; Dayal 2012).

[19] In functional incorporation, extension is obtained by means of perception, instead of in interaction with *objects* of perception. The difference was briefly discussed in the second chapter (Sect. 2.2.2.) when comparing differences between dispositional and occurrent EM.

conceptual pair is used in both philosophical (starting from Merleau-Ponty [1945]) and scientific (starting from Head and Holmes [1911]) discourses.

Body image refers to the intentional and reflective stance we have towards our bodies—the body experienced as an object. It is "composed of a system of experiences, attitudes, and beliefs where the object of such intentional states is one's own body" (Gallagher and Zahavi 2008, 164). *Body schema*, on the other hand, refers to the body as a pre-reflective subject of experiences—the body from a first-person perspective. It is a "system of sensory-motor capacities that function without awareness or the necessity of perceptual monitoring" (Gallagher 2005, 24). Body schema is our point of view to the world that is below conscious monitoring. We can grasp an apple without considering consciously the posture of our hands, jump over a small brook without considering the movements of our legs, and bend down to enter a car without considering how to adjust our posture. The objects that matter are the apple, the brook, and the car—not the body. If we hit our head while climbing into a car, we might lose the transparent stance and start seeing the body as an object again.

Even though these concepts are sometimes treated inconsistently in the literature,[20] it is undeniable that this distinction is legitimate, and it also has empirical support, namely in cases of ***double dissociation***. There is neuroscientific evidence of cases where the body image is injured but the body schema is intact, and the other way around, where the body schema is injured but the body image is intact. The best-known phenomenon is probably phantom limb pain—a patient who has lost a limb can still feel pain in "it" after amputation.

Gallagher (2005, 25) gives examples of both kinds. An example of the former (intact body schema, injured body image) is unilateral neglect, a condition following brain damage, where the patient fails to recognise items that are located on one side of a space (left or right). According to a research set-up by Denny-Brown, a person suffering from a brain lesion (in the hemisphere opposite of the neglected side) fails to be aware of the left side of her body. She doesn't dress the left side of her body or comb the left side of her hair. However, she is still able to use her left side quite

[20] Perhaps the most unfortunate confusion is that Merlau-Ponty's term *schéma corporel* was translated *body image* in the first English translation in 1962. Besides terminological confusions, there are also conceptual open questions—the case is far from settled, but after all, the distinction is more useful than confusing.

normally for motor tasks that require both sides of the body, such as using her left hand to button up clothes on the right side of her body (Denny-Brown et al. 1952, reported in Gallagher 2005, 25, 40).

An example of the latter (injured body schema, intact body image) is found in some rare cases of deafferentation. The symptoms are caused by a nerve fibre neuropathy that causes a loss of tactile sensations from the neck down. The patients can control their movements only by "cognitive intervention", for example, they use a visual perception of their body in order to compensate for the lack of proprioceptive and tactile feeling in their body (Gallagher 2005, 25; Cole 1995). Gallagher and Cole (1995) suggest that in some of these cases, the patient has lost the body schema, even though the body image stays intact.

An interesting aspect of body schema is that it is not tied to physiological body boundaries; instead it is plastic and open to adaptation. By *functional incorporation*, I refer to situations where the body schema takes external elements into it, namely, the body schema extends outside the biological body boundaries. To find out whether there is functional incorporation, we need to ask whether an external element can become incorporated into the *organisation of the system*, rather than merely extending an individual experiential state every now and then. When properly functionally incorporated, the previously external element becomes part of the organisation of an experiential system. Apart from enactivism, it is related to the *embodiment* thesis and to the Merleau-Pontian divide between the *objective, material body* and the *phenomenological, lived body*. Coupling happens with the latter kind of understanding of the body. It has an impact on the structure of experience, and the world is experienced through the incorporated external part transparently. Functional incorporation creates a *sense of ownership* over the incorporated part, which can be tested by removing the part: removing a prosthetic limb or a blind man's cane is closer to removing a body part than removing a fork or a notebook.

Let us then look at the term *functional*. Its purpose in my taxonomy is to differentiate functional from prosthetic incorporation. Colombetti (2016) draws from Merleau-Pontian differentiation and talks about 'phenomenological incorporation'. The reason why I don't follow her terminology, but instead talk about *functional* incorporation, is that firstly, according to my view, the phenomenological side is already present in the first two types (i.e. in momentary extension and integration), and secondly, I want to emphasise the importance of *functionality* as a purposeful, goal-driven property as opposed to the *phenomenological* side that would instead sound as if something was more passively received. (Of course, functional incorporation *is* phenomenological.) The external element needs to be under the "functional" control of the subject—controlled sensorimotorically by the active agent. Being a physical part (coupled within the biological body) is not sufficient.

Merleau-Ponty's (1945, 165–166) classic examples of incorporation[21] are famous and often cited in the literature. The best known is the blind person's cane: the cane is not (primarily) an object but functions as if it were a sense organ through which the blind person experiences her surroundings. Merleau-Ponty's other examples include a driver and her car, and a woman and her feather hat. The driver and the hatted woman have extended their bodily frontiers to include the size of a car and the feather, and they know (i.e. feel) without calculating whether they can pass through a narrow opening—just as we don't have to calculate our body size to find out whether we can pass through a door or not. "To get used to a hat, a car or a stick is to be transplanted into them, or conversely, to incorporate them into the bulk of our own body" (Merleau-Ponty 1945, 166).

From Merleau-Ponty's examples, the blind person's cane fits well with my definition of functional incorporation. The cane clearly fulfils a sensorimotor function, is transparent (or quasi-transparent), and extends the body schema. The hat and the car examples undoubtedly extend the body schema as well—but it is less clear whether they have a sensorimotor function or purpose of the right kind. A sensorimotor stance is certainly included, but it occurs as a side effect. Especially having a feather in one's hat would rather belong better to my definition of prosthetic incorporation. The function of the feather is not to extend its carrier's sensorimotor area or the scope of sensitivity (e.g. the feather is not comparable to a cat's whiskers). Even though something of this sort can happen when wearing a hat (we could call it a weak imitation of a cat's whiskers), feather hats are not usually used for sensorimotor probing. (Same goes for scarfs, wigs, facemasks, etc.—this is why I left out prosthetic extension from ECM in the previous section.)

The car example comes closer to functional incorporation, at least by changing the focus of interest a little. Here, too, the right sort of incorporation is of course technically also a side effect or by-product, since the

[21] Regarding the distinction presented earlier in this chapter between *habit incorporation* and *object incorporation* (see Colombetti 2016), in these examples, the notion of incorporation used by Merleau-Ponty (1945) goes of course under *object incorporation*.

main function of the car is—obviously—to enable passengers to move (faster than by foot) in space. But consider somebody driving on an oily or icy surface, or driving over small obstacles like stones. The feeling is probably even stronger with a sport car, since there is less material between the driver and the road. The focal point of sensorimotor sensitivity is transformed from the borders of the body to the borders of the car. In all of Merleau-Ponty's examples, pieces of the environment become incorporated into the body schema.

An external piece is properly functionally incorporated when it functions *transparently* as part of the system. Thompson and Stapleton (2009) call this requirement a *transparency constraint*, according to which: "*For anything external to the body's boundary to count as a part of the cognitive system it must function transparently in the body's sense-making interactions with the environment*". The transparency constraint is a phenomenological criterion for incorporation. It emphasises phenomenological considerations—the first-person experience of the subject—as a criterion for incorporation. When some element is incorporated into the body, it is no longer felt as an object, but the world is experienced through it. It is reminiscent of the transparent status of the biological body in normal experience, and it modifies the way in which world is experienced. Moreover, it doesn't stay in the phenomenological realm alone. I agree with Thompson and Stapleton (2009) when they hypothesise that "tools and aids that conform to transparency are incorporated into the neurophysiological body schema". This idea not only is based on philosophical hypothesising but also has empirical support (e.g. Maravita and Iriki 2004).

However, I think we should adjust the transparency constraint a little. The tool doesn't have to be absolutely transparent because, strictly speaking, it might be that no tool is completely transparent—otherwise it would no longer be a tool. A sufficient level of transparency is needed, so that the focus of the attention and significance is not in the tool, but in the target of the coupled system formed by the subject and her tool. Although not a physical tool, speaking a foreign language that one has learned as an adult is an example of this kind of partial transparency. At least when the speaker is not fluent in the language, there is always a certain barrier that prevents total transparency (as opposed to one's native language). Colombetti (2016, 240) describes this kind of partially transparent tool as "quasi-transparent, i.e., as 'withdrawn' from experience (not attended, not taken as an intentional object) but still present".[22] She continues that the tool is presented to its user as "that through which" the world is experienced. This "that through which" nature of the tool is necessary for functional incorporation.

When asking which tools can become functionally incorporated, one should not look at the tool itself but rather at its use. Whether a tool can be incorporated depends on the subject's interactive relation with it—rather than on the tool's design, internal structure, effective functioning, portability, and so on. Instead, it is crucial that the tool can be manipulated and controlled by the subject.

> It is as if the tool withdraws into the sensorium of the body, or rather, as if the tool extends not only the motor body, but also the sensorium of the body. This results into a feeling of quasi-transparency in which the tool co-enables our experience but is itself not in the focus of our experience. The quasi-transparency of the tool is related to a feeling of non-mediation, caused by the altered correlation between somatosensory and visual information. (De Preester 2011, 123)

Sometimes it is difficult to see whether a tool is a mediating or non-mediating tool, and placing it into one single group is complicated. For example, an artificial limb is an ambiguous case.[23] Consider a person using an artificial leg, for instance. Is it functionally or prosthetically incorporated? The answer depends on what dimension we are looking at, and how profoundly the device has been taken into the system. It goes without saying that the actual users of artificial limbs would prefer them to become as incorporated as possible. However, this doesn't always happen, for the artificial limb may remain an external tool that is not felt as part of the body. De Preester and Tsakiris (2009, 310) suggest that it should become a "*knowing* body-part, in other words, something that shares in the knowledge of the body". Experiments with artificial limb users have shown that they may become "knowing body-parts" but only in a limited sense (De Preester and Tsakiris 2009; see also Murray 2004).

[22] For quasi-transparency or semi-transparency, see also Ihde (1990) and De Preester (2011).

[23] Artificial limbs are of course called *prostheses*, but I use the word 'prosthesis' referring to prosthetic incorporation only, and artificial limb may or may not belong to that group regardless of this terminological matter.

Of course, there are exemplars of successful artificial limb incorporation, as, for example, the famous South-African sprinter Oscar Pistorius proves. Both of his legs were amputated in his infancy, and he wears running blades that enabled him to succeed not only in the Paralympics but also in the Olympics. The blades turned out to outperform biological legs in certain aspects that affected the running speed. The way Pistorius ran with his blades is an interesting case for science in many ways, and importantly for our purposes, it looks like a possible case of functional incorporation.[24] Hence, an artificial limb may or may not belong to the group of functional incorporation, depending on the case.

The benefits of restricting ECM to cases of functional incorporation are that it gives a demarcation criteria for ECM, and it ties 'embodied' (in*corporated*), 'enacted' (active, functional sensorimotor stance), and 'extended' (non-biological object/tool is included) together. The disadvantage is that it excludes cases that belong to 'momentary extension' and 'integrated extension' sections. Even though they clearly have an experiential side to them, the reason for leaving them out is that the coupling is not strong and stable enough—they only extend individual experiential states but don't become part of the organisation of the experiential system or subject. They don't contain a sense of ownership in the same way as functional incorporation does. However, this doesn't mean that extension based on momentary extension and integration would be non-experiential. As I have stated earlier, we don't transform into Chalmers's zombies in real-life cases of sensorimotor tasks and cognitive reasoning. However, these cases are exempted from my narrow definition of ECM. They can be taken rather as support for a general tendency for experiential extension.

This section aimed at differentiating functional incorporation from weaker sorts of extensions. However, as the example of an artificial limb showed, whether something counts as functional incorporation needs to be analysed case by case. In the following sections, I will investigate functional incorporation from slightly different angles. First, I will set the criteria for it even more explicitly in the form of *glue and trust* conditions for ECM, and in the final part of this chapter, I will discuss one concrete example of functional incorporation, namely *sensory substitution* technology.

[24] Although he doesn't run anymore, because he was convicted of killing his girlfriend in 2014 (Lindstrom 2019).

4.2.5 Summing Up the Four Degrees of Extension

	Functional or not?	*Incorporated or not?*
Momentary extension	No	No
Integrated extension	Yes	No
Prosthetic incorporation	No	Yes
Functional incorporation	Yes	Yes

This illustration sums up the crucial differences within the taxonomy I have outlined. Let us review briefly how the different levels of extensions are separated from each other. The difference between *momentary extension* and *integrated extension* is that the former is more fleeting—a tool that one uses as an external prop, for example, for a motor or cognitive task, but it is not taken into the system as a genuine part. Many of the tools we use every day belong to this group, such as knives, forks, pens, and toothbrushes. However, becoming *part of the system* is required for something to count as integrated extension. I suggested that Clark and Chalmers's original Otto example belongs to this group. Also, someone who has played, say, the violin long enough has probably *integrated* the instrument in a more stable way into her system. The first two groups of extensions, hence, are on the same scale, the latter just being a stronger and more permanent relation.

The third group of extension, *prosthetic incorporation*, is an instance neither of EM nor of ECM. It is not on the same scale of extensions with the other three groups but is included in the taxonomy in order that its difference from functional incorporation becomes evident. Prosthetic incorporation differs from *functional incorporation* in that the former doesn't involve sensorimotor control over the incorporated part—and the other exemplary features of functional incorporation remain absent as well. An example of a prosthetically incorporated part could be a cardiac pacemaker or an electronic neuron, whereas the best examples of functional incorporation are devices that have an active role in forming sensorimotor, perceptual experiences, such as a blind person's cane or a sensory substitution device.

Further, momentary and integrated extensions differ from functional incorporation in that the former pair do not involve a bodily sense of ownership that shapes the body schema, whereas functional incorporation does. When one loses a part that has been functionally incorporated (such as a blind person's cane), the loss is different compared to losing a tool that might have had an important status but which hasn't been part of the bodily self. Even in a momentarily extended or integrated tool use (such as using a pair of scissors or a paintbrush), the tool can be under the motor control of the user, and hence a feeling of control over the tool can develop. However, the relationship to the tool is still not as if it was part of the body or bodily functions. Removing the external part is a way to reveal whether a sense of bodily ownership was involved. Removing a cane or a *tactile visual sensory substitution system* (TVSS) device from a blind person is comparable to losing a body part, whereas taking away a paintbrush from an artist is not—even though the paintbrush might be of great importance to the artist. After having demarcated these degrees of extensions, let us now turn to the criteria that are needed specifically for ECM.

4.3 "Glue & Trust" Conditions

Glue and trust conditions refer to criteria that need to be fulfilled for there to be genuine extension. They are the guiding instructions when some external element is "glued" to the system tightly enough that it can be trusted as part of that system—so that a new coupled system is formed. The criteria might vary from case to case, and hence it is impossible to give all-encompassing rules that would fit all situations. The criteria nevertheless help us to demarcate genuine extension (or *integration*, in the case of EM, and *functional incorporation*, in the case of ECM). Setting the criteria is an effective way to resist the slippery slope of the *cognitive bloat* argument (which has already been discussed in Chap. 3). Let us start with how the limits of EM have been set in the literature and then compare how these criteria fit with ECM. Finally, we can then set the glue and trust conditions for ECM.

4.3.1 Glue & Trust Conditions for EM

Clark and Chalmers (1998) and Clark (2010) have phrased quite clearly their conditions for the extension of cognition. They stress, however, that these are not categorical conditions for extension but rather guidelines for what kind of features should be included at least in cases that resemble the Otto case. Clark and Chalmers's *glue and trust* conditions include:

1. That the resource be reliably available and typically invoked. [...]
2. That any information thus retrieved be more or less automatically endorsed. It should not usually be subject to critical scrutiny (unlike the opinions of other people, for example). [...]
3. That information contained in the resource should be easily accessible as and when required.

(Clark 2010, 46)[25]

The tool needs to be "glued" to the system: it needs to be a constant, reliable resource. When properly glued, it will be *trusted*: it is automatically accepted and not questioned.[26] The result is a *coupled system* in a two-way interaction—from the external resource to cognition and from the cognitive function to the resource. It is clear that these conditions are written with Otto and his notebook in mind. As said, they are set up using a *case-by-case method*, which makes them *a posteriori* rather than *a priori*. For example, it is clear that the conditions cannot be the same for situations where one is using a notebook compared to, say, a situation where one is using a prosthetic limb or a walking stick. According to Hurley (2010, 126), we should "consider externalist explanations on their empirical merits, case by case". For example, in the case of Otto and his notebook, an externalist explanation is clearly better than an internalist one, as the original text aptly shows. I consider the case-by-case method to be the best way to embrace glue and trust conditions (whatever they may be) because we cannot set *a priori* rules to match all situations, and that was never the aim of Clark and Chalmers's criteria.

As a more general condition, Clark and Chalmers originally offered the *parity principle*, according to which extended cognitive processes must play the same functional roles as brain-bound cognitive processes. The parity principle would serve as a convenient general rule (as opposed to a case-by-case rule) only if it wasn't so problematic when taken literally and not taken as an intuition pump (for criticism of the parity principle, see Chaps. 1 and 2). A better way to explain extension is the *complementarity argument*, according to which environmental functions can play different,

[25] Clark and Chalmers (1998, 17) added a tentative fourth requirement, *prior endorsement*: the information in the notebook must have been consciously endorsed at some point in the past. However, they themselves admitted that the fourth condition is dubious because not even all biological memories are consciously endorsed (consider, e.g. memory tampering).

[26] Clark and Chalmers never specified a literal meaning with the term 'glue & trust' (and in fact it is often referred to as 'trust & glue').

complementary roles from the internal, biological ones. However, it doesn't provide such an explicit toolkit we are after here.

Kim Sterelny (2010) is along the same lines with Hurley.[27] Instead of an all-embracing requirement for a trusted, reliable relationship between one agent and an external resource (such as Otto's relationship with his notebook), he stresses variability between different cases. However, he names three dimensions where we should go looking for clues: (1) the level of *trust* when a person uses an external resource, (2) the level of how much the external resource is *entrenched* and *individualised*, and (3) the level of how it can be *shared* with other agents.[28] Thus, Sterelny's way of demarcation is rather setting a scale than positing either–or responses. This way resembles the scale or spectrum of extensions I framed in the beginning of this chapter.

Further, Mark Rowlands (2009) suggests *ownership* as a demarcation criterion (for cognitive states in general, whether extended or not). "EM has an obligation to demonstrate the sense in which the extended processes it invokes—manipulation, transformation, and exploitation of relevant environmental structures—can be owned, or in an appropriate way belong to, a cognitive subject" (Rowlands 2009, 15). Thus, a demarcation criterion is not any material barrier but the belonging relationship. Gallagher (2013) trusts in ownership as well: the enactive engagements determine the ownership of a process. A similar requirement has been proposed by Mason Cash (2013, 65). According to his *responsibility condition*, the solution is "in our shared, evolving, normative and social practices of holding people responsible for their actions". Instead of asking "what makes a process cognitive", we should ask "what makes any thought or action mine". Likewise, Tom Roberts (2012) stresses responsibility: "Extended resources can house an individual's beliefs, I propose, only if she has taken responsibility for their sources in a suitable way". I think both *ownership* and the closely related *responsibility* are important aspects. Otto's notebook, for example, has to be under his control, "owned" by

[27] Sterelny represents an interesting position. He criticises Clark's EM for the reason that it is restricting the phenomenon of cognitive extension to a too narrow area, to specific cases such as the use of a notebook. Instead, he supports a broader view he calls *scaffolded mind*, where distinction between causal and constitutive relations with the environmental resources is not made.

[28] First two of his conditions are similar with Clark and Chalmers's conditions (just in slightly different terms), but the third condition about intersubjective dimension is new and missing from most of the criteria presented in the literature.

him, and hence he is responsible for the contents in it and for his actions based on the contents (as opposed to contents, e.g., in Wikipedia). I will soon evaluate how these conditions fit with the preconditions for ECM.[29]

Yet another proposal to set the limits is the *mutual manipulation* criterion that was discussed at length in the previous chapter. Apart from the original formulation by Craver (2007a) and the ensuing literature in the new mechanist approach, a similar account has been suggested by the supporters of EM, namely the *continuous reciprocal causation criteria* (Clark 2008; Palermos 2014). Its strengths and weaknesses were presented in the previous chapter. In what follows, I will evaluate whether these criteria help us to set the glue and trust criteria for ECM.

4.3.2 Criteria Compared with ECM

Let us now turn to ECM and reflect the above listed suggestions for demarcating EM: could they be used as demarcation criteria also for ECM? First, let us compare Clark and Chalmers's original glue & trust conditions with ECM. Should the externalisations of experiential states be reliable, available, automatically endorsed, and easily accessible?

Let's say a blind person's cane is a part of the material basis for his tactile experience. Is it necessary that the cane is constantly and reliably available for him? Wouldn't just one successful case of extension be enough to show that ECM is true? Of course, the same question can also be asked in the case of extended cognition. If Otto uses his notebook only occasionally (but successfully in those cases), or if the notebook is not easily accessible for him, and he needs a little effort to reach it, why should these mitigations affect EM? Does his notebook still count as reliable if he showers without it? On the other hand, Inga's memory is not always reliable either—for example, when she sleeps or is intoxicated. This shows how setting limits is always somewhat arbitrary. It is impossible to set absolute criteria for easy accessibility or reliability—they are always a matter of degree.

The answer to these questions lies in the scale of extensions I presented earlier. Even though one successful use of a blind person's cane or a notebook contains some kind of experiential extension (in the case of the blind person's cane) and some kind of cognitive extension (in the case of the

[29] I consider it promising that many writers have ended up with quite similar criteria (even though slightly different terms are used).

notebook), they would belong to the lower groups of extension, such as momentary *extension*. In order for Otto's use of a notebook to belong to the group of *integrated* extension, and in order for a blind person's use of a cane to belong to a group of *functional incorporation*, Clark and Chalmers's original conditions are required. Thus, it seems Clark and Chalmers were on the right track with their criteria—they clearly capture some essential features of what differentiates EM from the more short-term fleeting extension of individual cognitive states or processes.

I suggest the same applies to ECM—especially as a means to differentiate momentary extension from functional incorporation. One successful use of a blind person's cane or using a stick to reach for something on the highest shelf only extends an individual experiential state. However, for an experiential *system* to be extended, a more profound and stable relation is needed. Therefore, *availability*, *reliability*, and *accessibility* should also be counted in ECM's demarcation criteria. Further, with the blind person's cane, *automatic endorsement* comes along if the reliability and availability conditions are satisfied. When a person successfully uses a device, it becomes transparent in a way that we could say it is "automatically endorsed". Thus, this requirement should also be counted in the glue and trusts of ECM. However, I don't use these exact same terms for my criteria of ECM because my emphasis is different, and Clark and Chalmers's criteria fit into some wider concepts I use, such as *transparency* and *ownership*.

What about the demarcation suggestions set forth by others? I think most of the above-mentioned criteria of EM (especially ownership, responsibility, trust, entrenchment, and individualisation) apply more or less to ECM as well—the emphasis just varies slightly. It is clear that the *parity argument* is not a good criterion for ECM, but as I have made clear, it is not a good criterion for EM either. The requirements based on neither *mutual manipulability* nor *continuous reciprocal causation* are unproblematic—but again, the problems are not specific to ECM but apply to EM as well.

Consequently, are there no differences in the glue and trust criteria between EM and ECM? Even though all of the criteria for EM also count as criteria for ECM (at least with a slightly modified emphasis), the same doesn't apply the other way around—hence the criteria are not the same. Some of ECM's criteria apply to perceptual experience only, namely those conditions that deal with sensorimotor control and bodily stance. Thus, ECM requires more than EM.

4.3.3 Glue & Trust Conditions for ECM

In this section, I will outline the conditions that need to be fulfilled in order to have an instance of ECM. As we saw, Clark and Chalmers's original criteria with the supplementing conditions presented by others serve as a starting point for criteria for ECM. However, apart from these conditions, the core of my glue & trust conditions for ECM is the fourth type of extension presented in the taxonomy above, namely *functional incorporation*. With that notion, I refer to quite a wide-ranging phenomenon, as already stated earlier. The relevant features embedded in that notion are the following:

1. It needs to be part of the quasi-transparent bodily point of view
2. It needs to be under sensorimotor control of the subject
3. Ownership (entrenched, individualisation, and fluency)
4. Sense of agency
5. Sense-making attitude (a purpose, goal-drivenness)

These are the glue & trust conditions for ECM. They are not isolated from each other but are linked in several ways. In order to gain (quasi-) transparency, the external part needs to be taken under one's control—to make it one's *own* by individualising and entrenching it.[30] A feeling of ownership emerges gradually with constant practice and use. This creates fluency, which increases transparency: the tool becomes transparent for a fluent tool user. Because ECM deals with perceptual experience, the component needs to be under the bodily, sensorimotor control of the subject. When it is taken under the user's voluntary control, a sense of agency will be gained. All this requires a *sense-making* attitude—a programmed robot just following rules would not count as having conscious experiences.

Further, my suggestions for conditions are not in contradiction with Clark and Chalmers's criteria, but as said earlier, their original glue and trusts fit mine. I also consider that Sterelny's (2010) way of talking about dimensions and levels of extensions is an advantageous way to proceed. In a similar vein as with the scale of extensions I presented at the beginning of this chapter, the glue and trust criteria for ECM should not be taken as

[30] The ownership doesn't have to be restricted to the subject but can be shared between other individuals. Collaborative processes, shared tools, or environments that are 'owned' by a group of people can also contain ownership in the right sense (see Sterelny 2010; Cash 2013).

categorical but rather as a matter of degree. They provide a guideline of what should be taken into account—but still each case has to be examined separately, and the emphasis of the criteria might vary depending on the case. I do not think the lack of a clear-cut assemblage of all-embracing criteria should be seen as a problem for externalist theories because a need for categorical criteria stems from the idea that there is a distinct *mark of the cognitive* or *mark of the consciousness* at play. However, as I have argued earlier, the request for the mark of the mental is a misguided way of proceeding in demarcating cognition and consciousness.

To summarise, according to ECM, *the constitutive material components (i.e. the material basis) of some experiential states can include external elements on the condition that they are functionally incorporated into the system*. Functional incorporation involves *quasi-transparency*, *sensorimotor control*, *ownership*, *a sense of agency*, and *a sense-making attitude*. In the following section, I will discuss an example where the criteria are fulfilled.

4.4 An Example of Functional Incorporation: Sensory Substitution

Sensory substitution (SS for short) provides an illustrative and paradigmatic example of functional incorporation, that is, an example where all the glue & trust conditions of ECM are met. An increasing number of philosophers of perception, cognitive sciences, and neurosciences have been interested in this phenomenon—which is not surprising because it offers intriguing empirical ways to investigate some of the key questions in these research fields (see, e.g. a recent collection edited by Fiona Macpherson 2018). However, the relationship between sensory substitution and 4E-views has not been analysed thoroughly in the literature. There are some exceptions. In particular, Andy Clark has argued that SS provides empirical evidence for EM (e.g. Clark 2003, 2008; Kiverstein et al. 2015) and Alva Noë has argued that it gives empirical evidence for ECM (e.g. Hurley and Noë 2003; Noë 2004, 2009; see also Kiverstein and Farina 2012). In this section, I will outline the main features of sensory substitution, introduce some examples of SS-technology, and argue that the previously set glue & trust conditions are fulfilled in a proper use of SS-devices (after sufficient training).

The term sensory substitution refers to cases where one sensory modality is replaced with another sensory modality (e.g. vision with touch) or within the same sensory modality (e.g. touch in the fingers with touch in the forehead). Sensory substitution systems are used to restore or compensate for lost senses (e.g. for blind or deaf subjects) or to enhance or create completely new senses (e.g. a feeling of magnetic North)[31] by exploiting other senses. The latter is sometimes more specifically referred to as *sensory augmentation*. Users of SS-technology gradually learn, for example, to do localisation tasks, motor tasks such as batting a ball, and recognising the shapes of objects (Macpherson 2018; Bach-y-Rita 2002; Auvray and Myin 2009; Kiverstein et al. 2015).

Sensory substitution and augmentation are both forms of intervening sensory paths using an auxiliary system. It is possible because of brain plasticity: the devices exploit the cross-modal plasticity of the sensory cortex.[32] Instead of a biological sense organ, an artificial receptor is delivering the environmental information to the brain via a *human-machine interface*. Importantly, the device needs to be under the *motor control* of the user, for example, a hand-held camera that the agent can move. Especially the pioneering work of an American neuroscientist Paul Bach-y-Rita and his work in neuroplasticity and earliest forms of sensory substitution devices have opened up this field of research (and industry). (See, e.g. Bach-y-Rita, 2002; Bach-y-Rita and Kercel 2003; Auvray and Myin 2009; Macpherson 2011; Farina 2013.)

On a general note, sensory substitution is a reminder that the way we perceive the world is not the only possible way—it is only our take on it. The world appears very different to a hound dog, who mostly relies on olfactory sense, or to a bat, who uses echolocation to navigate in the dark. Besides helping to restore and compensate for lost senses, sensory substitution technology can also be used as way of accessing perceptual experiences that have previously been inaccessible for humans. Broadly understood, the phenomenon is not restricted to modern technology but

[31] The *feelSpace* project at the University of Osnabrück is conducting research on a sensory augmentation device: *naviBelt*, a vibrotactile magnetic compass belt that subjects wear around their waist. After training, the subjects learn to recognise magnetic North instead of just feeling a vibration at their waist. (See König et al. 2016; and for the project, https://www.feelspace.de/?lang=en).

[32] However, Kevin O'Regan (2018) has argued that the reason why sensory substitution is possible is actually not cortical plasticity as such but rather the sensorimotor changes in the modes of environmental interaction that the user undergoes while using the device. I am sympathetic to his claim because, as he says, stressing the role of cortical plasticity may lead to an assumption that the cortical areas are the only material realisers of experience.

has been used long before the rise of recent technological developments. Braille language as used by blind people is sometimes labelled as the most successful sensory substitution system. In Braille, visual sensing (reading) is substituted by tactile sensing in the fingertips. In a similar vein, sign language used by people with hearing defect is a way to replace auditory senses with visual senses. Yet another obvious example is a blind person using a cane, where visual information is replaced by tactile information. Bach-y-Rita and Kercel (2003, 541) even suggest that *reading* could be considered the first SS-system—because reading doesn't occur naturally but is a way of substituting auditory modality (the spoken word) with visual modality. These examples reveal the very elementary function of sensory substitution. However, in what follows, I refer specifically to the technological aids (that are also used in clinical settings) that have been developed during the last 50 years or so starting from Bach-y-Rita's pioneering work in the late 1960s, instead of the more general capability of replacing senses with other senses.

The most common SS-devices belong to two kinds of systems: visual-to-tactile systems, which convert visual pictures to tactile patterns, and visual-to-auditory systems, which convert visual pictures to patterns of sounds (Farina 2013, 640; Auvray et al. 2007, 416). Perhaps the best-known example of the former is *tactile visual sensory substitution system* (TVSS) and its application *Tongue Display Unit* (TDU) (see Bach-y-Rita and Kercel 2003), and of the latter, *The vOICe* (see Meijer 1992).

TVSS was first invented by Bach-y-Rita and colleagues in the 1960s and 1970s, and several versions have been developed since then, for example, the *Brain Port* and the *AuxDeco Forehead Sensory Recognition System* (*FSRS*) (see Nau et al. 2015). TVSS has been described as a way for the blind "to see by touch". TVSS subjects have a receptor in their skin that takes input from a head-mounted or hand-held video camera (there are different versions of electrotactile and vibrotactile receptors, which can be located e.g. in the abdomen, the back, the fingertip—and the tongue has proved to be a particularly suitable place). In response to the camera input, the unit in the skin activates and gives tactile stimulation, which is first felt as it is—tactile, cutaneous stimulation—but after the subject has used the device for some time, the feeling starts to be transformed from tactile to something that resembles visual experience. Whether visual or not, subjects can act *as if* they had visual experiences. They report experiencing images in space rather than on the skin. They can make judgement based on visual information, such as depth estimations, perspective, and hand-eye coordination tasks (Bach-y-Rita and Kercel 2003).

The best-known visual-to-auditory SS-device is *The vOICe*, first developed by Meijer (1992).[33] It is a way to translate visual inputs into audible patterns—from video to audio. Objects that are high in the visual field produce high-pitched sounds, and lower objects produce lower-pitched sounds. Lateral location is related to the balance of the stereo sound and brightness with loudness, and so on. The *human–machine interface* is simpler in visual-to-auditory devices than in TVSS. *The vOICe* consists of three components: a video camera embedded in sunglasses, a smartphone running the algorithm, and stereo earbuds (Farina 2013, 640). Technology needed for *The vOICe* and other kinds of visual-to-auditory devices is very common nowadays, which is an asset compared to visual-to-tactile technology. As with TVSS, subjects who use *The vOICe* for a sufficient period of time will be able to perform tasks that usually require visual modality, such as recognising objects[34] (See Clark 2008, 173; Auvray et al. 2007).

Apart from visual-to-tactile and visual-to-auditory, there is also, for example, *tactile-to-tactile* and *tactile-to-vestibular* sensory substitution systems. An interesting application of tactile-to-tactile sensory substitution technology is the so-called *leprosy glove* (Bach-y-Rita and Kercel 2003). Subjects who have lost peripheral sensation (e.g. due to leprosy) can start feeling again with their fingertips—or rather, with the fingertips of the gloves. The glove has artificial contact sensors coupled to a forehead-mounted tactile disc-array. After training, the subjects reported experiencing "the data generated in the glove as if they were originating in the fingertips, ignoring the sensations in the forehead" (Bach-y-Rita and Kercel 2003, 541).

There are a number of other applications of SS-technology as well, but these examples suffice for understanding the relevant features for discussion about functional incorporation and ECM. But before analysing how these applications of sensory substitution fit into my glue and trusts, one clarification should be made. Many writers have addressed the question of what kind of experience the substituted experience is. The options are roughly as follows: (1) It stays in the substituting modality (e.g. tactile in

[33] Another example of a visual-to-auditory device is the software program *EyeMusic* that translates visual stimuli into sounds (Nau et al. 2015).

[34] *The vOICe* has been reported to function for both early-blind and blindfolded-sighted subjects (Auvray et al. 2007).

the case of TVSS). (2) After sufficient training, it switches to the new modality (e.g. visual in the case of TVSS). (3) It is neither of these, but a genuinely new kind of perception, or as Auvray and Myin (2009) phrase it, a "novel form of perceptual interaction with the environment" (see Farina 2013 for an overview).

However, this question, even though otherwise highly intriguing, is not essential for my purposes. For it doesn't matter to which modality the experience belongs, as long as there is a transformation in sensory experience with the device—and that much is evident. After all, it is good to notice that all (at least present-day) sensory substitution technology is incomplete—no one has ever *really* seen the whole rich scenery with all the finest details with a tongue display unit or with sounds (for an overview of the limitations, see Deroy and Auvray 2012). However, this doesn't undermine my argument: it suffices that the subjects using the devices are going through sensory experiences that enable them to perform tasks that they wouldn't be able to perform otherwise.

Let us now turn to the question how the use of sensory substitution devices counts as an instance of functional incorporation. The central feature in order for sensory substitution technology to work is "goal-driven motor engagement" (Clark 2007, 269). The subject needs training in order to get gradually used to the device, to gain a *sense of ownership* over it. In order to work, it is important that the "new agent-world circuits be trained and calibrated in the context of a whole agent engaged in world-directed (goal-driven) activity" (Clark 2007, 270). For example in TVSS and *The vOICe*, one needs to be in motor control of the camera; uncontrolled, passively received visual information coming through the camera will not produce an effective result. The training period can be seen as an individualisation process: the user makes the device her own, and it becomes rooted and entrenched in the system. Hence, the requirements of *sensorimotor control* and *sense of ownership* are fulfilled. These features also ensure that a *sense of agency* is gained. The user of the SS-device is in charge—she actively brings about the changes in her perceptual sphere.

Furthermore, there is an interesting similarity between sensory substitution and the enactivist notion of *sense-making*. According to enactivism, there is no pregiven world, as agents create their environments (or their *Umwelts*, to use Uexküll's term) with the perceptual machinery they have. The same environment appears different for a bat and for a bee, and for a blind and for a sighted person. Sensory substitution technology opens a door for experiencing other kinds of previously unfamiliar *Umwelts*—ways of making sense of the environment in new ways. For example, a blind person who gets access to what usually belongs to visual modality, or a person with normal human sensory modalities who gets access to a completely new modality—such as a feeling of magnetic North—gains access to an environment that is felt differently due to the coupling with the SS-device. The teleological aspects embedded in the notion of sense-making are also present in using SS-devices. Using SS-devices is always goal driven: the successful use requires "goal-driven motor engagement", as mentioned earlier. Thus, we can say that in the successful use of sensory substitution technology, there is a goal-driven, *Umwelt*-creating sense-making attitude at work.

Finally, the (quasi-)transparent bodily point of view develops with the successful use of SS-technology. For example, with TVSS, the subject stops feeling the tingling sensation on the skin and starts sensing the external cause (such as a ball being thrown towards her). The subject doesn't feel the device anymore but the world outside her, that is, the coupled user–device system. In the same way as a blind person doesn't feel the cane in her hand, but the world at the tip of her cane, a TVSS-user doesn't feel the device but the world around her. The device fades from view—in other words, it becomes transparent. Thus, the (quasi-) transparency constraint is also fulfilled. This new coupled system, or the "new agent-world circuit", emerges gradually when transformation from a mere tool to a (quasi)-transparent incorporated part of the perceptual machinery happens. The SS-device becomes part of the constitutive base for the cognitive and experiential system, letting the user of the device have experiences that would not otherwise be possible.

Thus, all of the glue and trust requirements are satisfied in the successful use of sensory substitution devices: the device becomes part of the quasi-transparent bodily point of view; it is under the sensorimotor control of the subject; a sense of ownership develops; a sense of agency develops; and a sense-making attitude is involved. Thus, a sensory substitution device (when properly used, and after sufficient training) becomes *functionally incorporated* into a subject's perceptual and experiential system. The result is a new integrated system, or a new systemic whole that consists of the subject and the device—and the constitutive basis of those new perceptual, sensory experiences is extended to include also the device.

Perhaps an opponent of ECM would admit some sort of extension in sensory capabilities but still be doubtful whether there is in fact *conscious experience* involved in substituted sensing. Bach-y-Rita (2002) explicitly asks whether there is *qualia* involved, and his answer is that there most likely is when the SS-system has been in use since early childhood. According to him (2002, 510), results from the use of SS-systems for blind babies indicate that when used from infancy, the device becomes a part of the child's *personality*, and a phenomenal feel comparable to the phenomenal feel of the sighted can develop.

However, I think Bach-y-Rita's question is misleading. Insofar as conscious experience is understood in the enactivist way, the question of whether there is some extra quality of experientiality or phenomenality involved (whether we call it *qualia* or *something-it-is-like-ness* or something else along these lines) simply doesn't arise. Successful use of a SS-system clearly involves conscious experience when conscious experience is understood as an active embodied sense-making interaction with one's environment—and this is exactly what using SS-devices is all about. This much should also be evident for someone who denies ECM. Further, I think the sceptic should also acknowledge that SS is a case of functional incorporation as I have described it—whether or not she agrees that functional incorporation is a sign of ECM. (However, if functional incorporation is accepted as a demarcation criterion for ECM, then we need to accept that sensory substitution is a case of ECM.)

Michael Wheeler (2015, 144–146) has taken up this sceptic view.[35] His aim is not to deny that SS counts as incorporation (or functional incorporation, in my terms), but he questions the inference from incorporation to the SS-device becoming part of the material realiser of the substituted experience. According to him, the jury is still out on the latter step:

> Let's agree that sensory substitution is a case of incorporation. Is incorporation sufficient for [ECM]? The answer, I think, is 'no'. It's simply hard to know why a modification to the internal representational resources of the subject—a modification to the neurally located machinery that determines how the agent categorizes portions of the world as body and not-body—should have the implication that the material vehicles where phenomenal conscious[ness] is realized now include the elements that have been newly represented as body rather than not-body. (Wheeler 2015, 145)

That is, Wheeler claims that the use of the SS-technology does change the body schema—what the agent counts as "body and not-body"—but

[35] His target is Kiverstein and Farina's (2012) argument that resembles mine, according to which sensory substitution is a case of incorporation, and hence a case of ECM.

that the change happens at the internal representational level. He continues (2015, 146) that what has changed is "the experience of where the bodily boundary is located", but says that what hasn't "thereby changed is where, in relation to the bodily boundary, experiences are located".

First, let us consider Wheeler's appeal to change in body schema as a change in internal representation of the body in the brain. According to him, even though what the agent counts as belonging to her body and not belonging to her body changes, it is only due to reconfigurations of the representations in the brain. Is the use of SS-technology actually just a way to modify the brain?

In his objection, Wheeler bypasses crucial elements from 4E-framework, namely, the constitutive nature spelled out in the *embodied* and *enacted* theses. Our embodied interactions and engagements in the environment cannot be reduced to representations in the brain without violating the core claims of these theories. Functional incorporation, whether actualised with a sensory substitution device, a blind cane, or some other instrument, shows how the tool can become part of the realising package. If experience *is* that dynamic interaction going on between the perceiver and the environmental elements, it cannot be shrunken to neural representations at the sub-personal level (as I have argued in the second chapter leaning on Ward 2012).

Denying the external part's contribution, and appealing to the brain-level representational resources, would be the same as to say that what actually matters in Otto's case when he looks at his notebook is the brain-level representations, and not the notebook per se. If one accepts Wheeler's reduction to the neural, representational level in the case of SS-technology, one has to do so with Otto as well and therefore deny EM.[36] However, the main message of Clark and others was precisely to add the notebook to the sub-personal explanation (without denying the brain-level activation).

Finally, let us consider the closing crystallisation of his criticism (quoted above) that states that *the experience of the bodily boundaries* has changed, but that *where experiences are realised* hasn't changed. This remark is of course just another way to repeat his earlier message, but this formulation reveals the gap he creates between the body schema and the core of the experiencing subject. As I have shown in this chapter and in the argument based on embodiment (presented in Chap. 2), if the body has leaked to

[36] Or if Wheeler would like to keep EM, he should present further arguments why the situation is different with consciousness in comparison with cognition.

the environment—or to put it the other way around: we have incorporated environmental elements into our bodies—"where experiences are realised" is (sometimes) extended or leaked as well.

If we accept my argument based on embodiment, Wheeler's criticism fails. However, as I mentioned (in Chap. 2), the conclusion of that argument doesn't follow logically from the premises, and hence it is possible to deny it (and Wheeler seems to join to that camp). One can, at least at a first glance, accept that (1) experience is embodied, that is, the body is (one of) the material realiser of experiences; (2) the body is sometimes extended to the environment with means of functional incorporation; and yet deny that (3) those incorporated parts sometimes are material realisers for certain experiences. However, this is a very narrow way to interpret the first and second premises. This reading becomes untenable if the embodiment of experiences and the capacity to functionally incorporate tools into our body schema are understood as constitutive claims. Wheeler's argument surely holds if he denies the thesis of embodied experience altogether. But if embodiment is taken seriously, it is a small step to the functionally incorporated tools.

4.5 Conclusions

In this chapter, I differentiated between four degrees of extension. Even though also the first two of them (*momentary* and *integrated extensions*) can involve experiential states, they are not stable enough to provide a relation that is needed for ECM. I argued that only the last one, *functional incorporation*, is suitable for ECM. For clarificatory purposes, I introduced an additional, separate group (*prosthetic incorporation*) that doesn't qualify as any kind of extension as extension is understood in EM and ECM.

I then set up the criteria needed for ECM, namely the *glue and trust* conditions. The external part needs to be (1) part of the (quasi)-transparent bodily point of view, (2) under the sensorimotor control of the subject, (3) entrenched and individualised as the subject's own, and there needs to be (4) a sense of agency involved, and (5) a goal-driven, sense-making attitude.

Finally, I analysed an example that fulfils these conditions. I showed that the use of sensory substitution devices, such as TVSS or *The vOICe*, fits well with my description of functional incorporation and the features included in it, namely the glue & trust criteria for ECM. I discussed an objection presented by Michael Wheeler, according to which sensory substitution may be a case of incorporation, but that it still doesn't prove that the experiential base would be extended. I aimed to show that his objection would only hold with a high price: one would also need to deny EM (or present further reasons why consciousness differs from cognition), and the embodiment and enacted theses as constitutive claims.

References

Adams, F. & Aizawa, K. (2008): *The Bounds of Cognition*. Oxford: Blackwell.

Atzori, P. & Woolford, K. (1995): "Extended-body: Interview with Stelarc". *Ctheory*.net.

Auvray, M., Hanneton, S., & O'Regan, K. (2007): "Learning to perceive with a visuo-auditory substitution system: Localisation and object recognition with The vOICe". *Perception* 36, 416–430.

Auvray M., & Myin E. (2009): "Perception with compensatory devices. From sensory substitution to sensorimotor extension". *Cogn Sci* 33: 1036–1058.

Bach-y-Rita, P. (2002): "Sensory Substitution and Qualia". In Noë A., & Thompson E. (eds): *Vision and Mind: Selected Readings in the Philosophy of Perception*, Cambridge, MA: MIT Press, 497–514.

Bach-y-Rita, P., & Kercel, S.W. (2003): "Sensory Substitution and the Human-Machine Interface". *Trends in Cognitive Sciences* 7: 12, 541–546.

Beck, B. B. (1980): *Animal Tool Behavior: The Use and Manufacture of Tools*. New York: Garland Press.

Berti, A., & Frassinetti, F. (2000): "When far becomes near: Re-mapping of space by tool use". *Journal of Cognitive Neuroscience* 12, 415–420.

Cash, M. (2013): "Cognition without borders: 'Third wave' socially distributed cognition and relational autonomy". *Cognitive Systems Research*, 25–26, 61–71.

Chalmers, D. (2019): "Extended Cognition and Extended Consciousness". In Colombo, M., Irvine, E., & Stapleton, M. (eds.), *Andy Clark and His Critics*. New York, NY: Oxford University Press.

Clark, A. (2003): *Natural-Born Cyborgs: Minds, Technologies, and the Future of Human Intelligence*. Oxford, UK: Oxford University Press.

Clark, A. (2007): "Re-Inventing Ourselves: The Plasticity of Embodiment, Sensing and Mind". *Journal of Medicine and Philosophy*, 32: 263–282.

Clark, A. (2008): *Supersizing the Mind: Embodiment, Action and Cognitive Extension*. Oxford, UK: Oxford University Press.

Clark, A. (2009): Letters. London Review of Books, 31(6). Retrieved from http://www.lrb.co.uk/v31/n06/letters.

Clark, A. (2010): "Memento's Revenge: The Extended Mind, Extended". In Menary, R. (ed.), *The Extended Mind*. Cambridge, MA: MIT Press, 43–66.

Clark, A. & Chalmers, D. (1998): "The Extended Mind". *Analysis* 58: 1, 7–19.

Cole, J. (1995): *Pride and a Daily Marathon*. Cambridge, Mass.: MIT; originally London: Duckworth, 1991.
Colombetti, G. (2016): "Affective Incorporation". In J.A. Simmons & J. E. Hackett (eds.): *Phenomenology for the Twenty-First Century*. Palgrave Macmillan.
Colombetti, G. & Roberts, T. (2014): "Extending the Extended Mind: The Case for Extended Affectivity". *Philosophical Studies* 172 (5): 1243–1263.
Craver, C. F. (2007a): *Explaining the Brain: Mechanisms and the Mosaic Unity of Neuroscience*. Oxford: Oxford University Press.
Dayal, G. (2012): "For Extreme Artist Stelarc, Body Mods Hint at Humans' Possible Future". *Wired*. 5.12.2012.
Denny-Brown, D., Meyer, J. S. & Horenstein, S. (1952): "The significance of perceptual rivalry resulting from parietal lesion". *Brain*, 75: 433–471.
Deroy, O., & Auvray, M. (2012): "Reading the world through the skin and ears: A new perspective on sensory substitution". *Frontiers in Psychology*. https://doi.org/10.3389/fpsyg.2012.00457
De Preester, H. (2011): "Technology and the Body: The (Im)Possibilities of Re-embodiment". *Foundations of Science*, 16: 119–137.
De Preester, H. & Tsakiris, M (2009): "Body-extension versus body-incorporation: Is there a need for a body-model?" *Phenomenology and the Cognitive Sciences*, 8(3): 307–319.
De Vignemont, F. (2018): "The Extended Body Hypothesis: Referred Sensations from Tools to Peripersonal Space". In Newen, A., Gallagher, S. & De Bruin, L. (Eds): *The Oxford Handbook of 4E Cognition*. Oxford: Oxford University Press.
Farina, M. (2013): "Neither touch nor vision: sensory substitution as artificial synaesthesia?" *Biol Philos*, 28: 639–655.
Farkas, K. (2012): "Two Versions of the Extended Mind Thesis". *Philosophia* 40(3): 435–447.
Gallagher, S. (2005): *How the Body Shapes the Mind*. New York: Oxford University Press.
Gallagher, S. (2013): "The socially extended mind". *Cognitive Systems Research*.
Gallagher, S. & Cole, J. (1995): "Body schema and body image in a deafferented subject". *Journal of Mind and Behaviour*, 16: 369–390.
Gallagher, S. & Zahavi, D. (2008): *The Phenomenological Mind*. Routledge.
Head, H., & Holmes, G. (1911): "Sensory disturbances from cerebral lesions". *Brain*. 34: 102–245.
Hurley, S. (2010): "The Varieties of Externalism". In Menary, R. (ed.), *The Extended Mind*. Cambridge, MA: MIT Press, 101–153.
Hurley, S. & Noë, A. (2003): "Neural Plasticity and Consciousness". *Biology and Philosophy*, 18(1), 131–168.
Ihde, D. (1990): *Technology and the lifeworld*. Bloomington: Indiana University Press.

Kirchhoff, M. D. & Kiverstein, J. (2019): *Extended Consciousness and Predictive Processing: A Third Wave View*. London, UK: Routledge.

Kirchhoff, M. D. & Kiverstein, J. (2020): "Attuning to the world: The diachronic constitution of the extended conscious mind". *Frontiers in Psychology*. 11:1966.

Kiverstein, J., Farina, M., & Clark, A. (2015): "Substituting the Senses". In M. Matthen (Ed.). *The Oxford Handbook of the Philosophy of Perception*, Oxford, UK: Oxford University Press, 659–678.

Kiverstein, J. & Farina, M. (2012): "Do Sensory Substitution Devices Extend the Conscious Mind?" In Paglieri, F. (ed.): *Consciousness in Interaction: The role of the natural and social context in shaping consciousness*, Amsterdam: John Benjamins, 19–40.

König, S. et al. (2016): "Learning new sensorimotor contingencies: Effects of long-term use of sensory augmentation on the brain and conscious perception". https://psycnet.apa.org/record/2017-01244-001

Lindstrom, S. (2019): "Oscar Pistorius". *Encyclopædia Britannica*, November 18, 2019. URL: https://www.britannica.com/biography/Oscar-Pistorius

Macpherson, F. (2011): "Individuating the senses". In Macpherson, F. (ed.) *The Senses: Classic and Contemporary Perspectives*, Oxford: Oxford University Press.

Macpherson, F. (ed.) (2018): Sensory Substitution and Augmentation. Proceedings of the British Academy, Oxford University Press.

Meijer, P. B. L. (1992): "An experimental system for auditory image representations". *IEEE Transactions on Biomedical Engineering*, 39(2), 112–121.

Maravita, A., & Iriki, A. (2004): "Tools for the body (schema)". *Trends in Cognitive Sciences*, 8(2), 79–86.

Menary, R. (2007): *Cognitive Integration: Mind and Cognition Unbounded*. New York: Palgrave Macmillan.

Menary, R. (2010): "Cognitive Integration and the Extended Mind". In Menary R (ed.) *The Extended Mind*. Cambridge, MA: MIT Press, 227–244.

Meraz N. S., et al. (2018): "Modification of body schema by use of extra robotic thumb." *Robomech Journal* 5. https://doi.org/10.1186/s40648-018-0100-3

Merleau-Ponty, M. (1945/1962): *Phenomenology of Perception*. Trans. Colin Smith 1962. London: Routledge Press.

Murray, C. D. (2004): "An interpretative phenomenological analysis of the embodiment of artificial limbs". *Disability and Rehabilitation*, 26(16), 963–973.

Nau, A., Murphy, M., & Chan, K. (2015): "Use of sensory substitution devices as a model system for investigating cross-modal neuroplasticity in humans". *Neural Regeneration Research*, 10(11), 1717–1719.

Noë, A. (2004): *Action in Perception*. Cambridge, MA: MIT Press.

Noë, A. (2009): *Out of Our Heads*. New York: Hill & Wang.

Noë, A. (2015): *Strange Tools: Art and Human Nature*. New York: Hill & Wang.

O'Regan, K. (2018): "Rewired animals and sensory substitution: the cause is not cortical plasticity". In Macpherson, F. (ed.) *Sensory Substitution and Augmentation*. Proceedings of the British Academy, Oxford University Press.
Palermos, S. (2014): "Loops, constitution, and cognitive extension". *Cognitive Systems Research*, 27, 25–41.
Roberts, T. (2012): "Taking responsibility for cognitive extension". *Philosophical Psychology*, 25, 491–501.
Rowlands, M. (2009): "Extended cognition and the mark of the cognitive". *Philosophical Psychology*, 22(1), 1–19.
Rupert, R. D. (2009): *Cognitive Systems and the Extended Mind*. Oxford: Oxford University Press.
Sterelny, K. (2010): "Minds: Extended or scaffolded?" *Phenomenology and the Cognitive Sciences*, 9, 465–481.
Szucs, A. P. et al. (2000): "Interacting biological and electronic neurons generate realistic oscillatory rhythms". *NeuroReport* 11: 1–7.
Thompson, E. & Stapleton, M. (2009): "Making Sense of Sense-Making: Reflections on Enactive and Extended Mind Theories". *Topoi* 28(1): 23–30.
Vold, K. (2015): "The parity argument for extended consciousness". *Journal of Consciousness Studies* 22: 16–33.
Ward, D. (2012): "Enjoying the spread: Conscious externalism reconsidered". *Mind*, 121(483), 731–751.
Wheeler, M. (2015): "Not What it's Like but Where it's Like: Phenomenal Consciousness, Sensory Substitution, and the Extended Mind". *Journal of Consciousness Studies*, 22(3–4): 129–147.

CHAPTER 5

Possible Objections to Extended Conscious Mind

5.1 Introduction

In this chapter, I discuss four objections or counter-arguments that have been set forth against ECM. They are (1) the argument from high bandwidth, (2) predictive processing, (3) the brain-in-a-vat thought experiment, and 4) dreaming and hallucinations. The reason I have chosen exactly these counter-arguments is that I think they are either the most challenging (1 & 2), or the most common (3), or the most interesting (4). I will answer them and conclude that none of them refutes ECM.

The first two counter-arguments (as they are used by Clark) aim to block the inference from EM to ECM, that is, my first argument in the second chapter. The last two criticisms especially challenge my second argument for ECM (the inference from sensorimotor enactivism to ECM) but also my third argument (the inference from embodiment to ECM). However, I will argue that they are not successful in cutting off the inference relations of these arguments, and hence none of them refutes ECM.

The last two objections touch upon the same concern: if we can have (or conceive, in the case of the brain-in-a-vat thought experiment) instances of phenomenal experience without access to the environmental interactions, how can we claim that the interaction has a constitutive role for (other, e.g. veridical, non-envatted fully embodied waking-life) phenomenal experiences. The concern indeed is real, of which the existence of

P. Telakivi, *Extending the Extended Mind*, New Directions in Philosophy and Cognitive Science,
https://doi.org/10.1007/978-3-031-35624-7_5

dreams is perhaps the clearest expression. Another real-life example is receiving electric impulses in one's brain and in this way creating phenomenal experiences. Moreover, there are a number of different brain injury-based pathologies, such as the locked-in syndrome, that are manifestations of the same phenomenon. The reason why I took the highly speculative brain-in-a-vat thought experiment alongside these empirical cases is that it puts into words the centuries-old scepticism about the constitutive role of the outside world for our perception and mind.

5.1.1 Clark's Criticisms of ECM

Because one person (Andy Clark) has come up with so many counter-arguments for ECM, I will start this chapter by briefly summing up his line of thought. The contradiction between Clark being on the one hand the first and most famous supporter of EM, and on the other hand the most ardent opponent of ECM, is a thrilling one. Of course, the discrepancy is not only thrilling but also inconsistent, as I have already presented in the first argument of the second chapter (where I argued that if we accept EM, we need to accept ECM as well), and as I will show in this chapter as well.

As was briefly mentioned in the third chapter, Clark (2009a) argues against ECM along the same lines as Adams and Aizawa did against Clark's own EM: he claims that the relation between the external tool and conscious experience is causal but not constitutive. Accordingly, Clark continues, ECM commits a causal–constitution (C–C) fallacy. However, Clark's approach or underlying method is different from Adams and Aizawa's. Whereas the latter's argument was drawn from metaphysical considerations, Clark's argument is supposed to be based on empirical grounds. Clark (2008, 138) himself admits that the metaphysical C–C debate is "unproductive". We could call Clark's version an *empirical C–C fallacy*. Whereas Adams and Aizawa appealed to the metaphysical assumption about the lack of the "mark of the cognitive", Clark appeals to empirical reasons, such as the superior processing power of the biological brain. Even though a fact like the processing power of the brain is undoubtedly a more empirically verifiable/falsifiable than Adams and Aizawa's non-derived content, I claim that it nevertheless falls into the same kind of *a priori* claim about the "mark of the conscious". It leans on a presupposed border between conscious experience and the rest. (I will return to this issue in the answers for the high-bandwidth argument.)

Of the four counter-arguments presented in this chapter, two are used by Clark, namely the arguments from high-bandwidth and predictive processing. The former states that high-speed processing power is crucial in the construction of conscious experiences, and only neural processes are able to achieve it (Clark 2009a). The latter states that the neuroscientific theory of predictive processing is able to explain all the positive features of ECM while securing internalism (Clark 2012). Note, however, that Clark does not think that predictive processing is in conflict with externalism in general. On the contrary, he declares that they have much in common and that predictive processing is a "perfect neuro-computational partner for work on the embodied mind" (Clark 2016). He simply considers that conscious experience is out of the scope.

However, Clark explicitly admits that two of the counter-arguments that I will present *do not* refute ECM, namely the brain-in-a-vat argument and the argument from dreams and hallucinations. Clark (2009a, 981) recapitulates neatly why the envatted brain argument is mistaken:

> If the vat does not fill in everything the world provides, the experiment is unfair. If it does, it cannot prove anything, as the filled in contributions might [...] still be essential for that very experience. The brain-in-a-vat considerations are thus unable to advance the argument.

Furthermore, according to Clark (2009a, 981), the existence of states such as dreams and hallucinations can only show that there are *some experiences* that are not realised by means of external interaction. This doesn't yet tell anything about the rest of the experiences. He also acknowledges that the cases where the body and the world are in the loop outnumber the non-loop cases.

These rejections are reasonable of course, since these counter-examples would threaten Clark's own EM as well. The first intuition is that the brain-in-a-vat thought experiment, if successful, would upset EM easily. It seems that Clark realises that if used against ECM, it would also rebut EM. The rejection of the dream counter-argument, however, is a slightly more perplexing case. Dream experiences are experiential states, and hence in the range of ECM and not EM—and for that reason one would think Clark would want to employ it. Nevertheless, in his 2009 paper, Clark is hesitant to use it against ECM because he acknowledges that ECM follows the same pattern as his own EM: for EM or ECM to be a legitimate theory, it is not required that all cognitive (in the case of EM) or all experiential (in the case of ECM) states be extended. This argumentation is the most straightforward reason to reject the counter-argument based on dreams and hallucinations.

However, as we will see in Sect. 5.5., the dream argument leads to some complexities and requires a more thorough treatment with finer details. Moreover, even though Clark rejected it in 2009, he re-adopted it in his 2012 paper, where he states that dreams and hallucinations are problematic for ECM. He has a solution: unlike ECM and enactivism, which are unable to explain these phenomena, a framework from cognitive science, that is, *predictive processing*, is capable of offering a coherent explanation for them. In what follows, I will show why Clark should also drop his counter-arguments based on high bandwidth and predictive processing—if he wishes to uphold EM.

5.2 The Argument from High Bandwidth

Clark (2009a) bases his first counter-argument against ECM on a short passage about timing and bandwidth from a foreword to his own (2008) book written by Chalmers.[1] Chalmers's suggestion offers Clark a concrete explanation how ECM falls into the above-mentioned empirical causal–constitution fallacy. Clark draws a picture according to which ECM fails because the material basis of consciousness requires access to information on a very high bandwidth, and our relation to the environment is on too

[1] Chalmers (2008) only throws this idea in the air; he does not develop it further. However, in his more recent (also short) treatment considering EM and ECM (Chalmers 2017), he admits that the high-bandwidth argument might have been wrong because some perceptual processes, such as vision, are extremely high in bandwidth. He alters the argument slightly: "I think the right explanation is not that consciousness requires high-bandwidth access to information, but that it requires relatively *direct* access" (Chalmers 2017, italics original). This way of reasoning is not very far from the original high-bandwidth claim and especially because Chalmers doesn't specify very clearly what means by "direct" (and if direct access is needed, why is it not needed in the case of occurrent cognitive states?). It is difficult to see how "direct" is actually different from temporally high (i.e. quick) bandwidth. However, this new counter-argument is in a sense more metaphysical than the high-bandwidth argument, since Chalmers's argumentation relies on considerations very similar to arguments based on the differentiation between derived vs. non-derived content (see Adams and Aizawa 2001). Moreover, Chalmers suddenly appeals to action–perception loops or sensorimotor interaction as a requirement for truly extended cognition—a requirement that has been promoted by sensorimotor enactivists and supporters of ECM all the time and which in no way would refute ECM but quite the contrary would promote it.

low a bandwidth. Clark develops Chalmers's idea and thus appeals to the biological brain as "the only adequate vehicle" for consciousness.

Thus, speed matters. Clark suggests that there are timescales that are only reachable for neural processes. According to him, consciousness "requires cortical operations that involve extremely precise temporal resolutions" (Clark 2009a, 985). He combines this with Eliasmith's (2008) suggestion that neural and extra-neural dynamics are different in the speed of information flow (bandwidth) and that the extra-neural body functions as a "low-pass filter". A low-pass filter is a physical barrier that allows low-frequency signals through and blocks high-frequency signals. For example, the walls of a room function as a low-pass filter for a sound: the lower frequencies (such as the base) are heard better than the higher frequencies outside the room (Clark 2009a, 985).

> [T]he extra-neural body, implicated in all cases of active vision and motor loops, acts as a kind of low pass filter for signals coming from the environment. What this means in practice is that for phenomena that depend on, for example, the very fast temporal binding or processing of signals, the only locus in which such operations can (as a matter of fact) occur lies within the brain/CNS. (Clark 2009a, 985)

Hence, conscious experiences cannot be extended because everything that involves bodily, sensorimotor actions (e.g. eye saccades) will be "screened off" by the bodily low-pass filter. Clark ends up noting that the bodily, environmental aspects *do have* a role in constructing conscious experiences, but the role is merely causal.

5.2.1 *Answers to the High-Bandwidth Argument*

This counter-argument is problematic in many ways. *First*, it presumes *a priori* predominance of the neural that is very much in conflict with Clark's larger views about the human mind. It draws a principled barrier between the brain and the rest of the body (and the world). As I wrote earlier, Clark's reasoning resembles Adams and Aizawa's original critique towards his own EM. High bandwidth is used as the *mark of the conscious* in a similar manner as non-derived content was used as the *mark of the cognitive*. Even though the argument is based on two empirical claims (information flow on a high bandwidth and body as a low-pass filter), the way they are used builds up a principled gap between the neural and non-neural, and consciousness and the rest, respectively. Moreover, it also seems the latter of these claims is false on empirical grounds, as we will soon see.

Second, the argument is based on dubious assumptions about the body as a low-pass filter that do not have empirical support. One problem in differentiating high and low bandwidth is that it is dependent on the feature we want to describe and cannot be defined independently. Vold (2015) has criticised Clark's high-bandwidth argument based on research on visual perception. Visual modality is a non-neural system and as such should be low in bandwidth if the high-bandwidth arguments were true. However, visual perception is actually higher in speed than the neural processes dealing with vision.

> Information about the surfaces of objects is transferred when light hits the eye, which is subsequently transmitted to the brain. But the brain, which cannot transfer or receive information at the speed of light, slows this information processing down. So non-neural processes must be constantly reporting information back to the brain, through the low-pass filter Clark describes, at least as quickly as neural processes can operate. (Vold 2015, 21)

Therefore, it seems that the claim that the body is a low-pass filter is simply mistaken. This alone suffices to refute the high-bandwidth counter-argument because even if the high-speed processing were still necessary, there is no reason why it should be delimited in neural processing any more. All in all, as an empirical claim, the high-bandwidth argument is very ambiguous since Clark does not specify which neural activity he refers to.

Third, the fact that conscious experiences require fast neural processing, even if true, only shows that neural processing is necessary (as a matter of contingent, empirical fact about humans, as Clark also reminds us) for consciousness. No one would deny this, and it is fully compatible with ECM. Neural processing is a necessary criterion for conscious experiencing, and it is undoubtedly high in bandwidth. However, the high-bandwidth argument gives no reasons why the neural sphere would be a sufficient condition for realising conscious experiences. The fact that the brain is necessary does not entail that the brain is sufficient for constructing experiences.

Fourth, even if it were true, it would also debunk occurrent EM (which is certainly an unwanted result for Clark). According to Vold, there is no evidence that high bandwidth would *not* be necessary for cognition. For example, motor skills in expert behaviour, such as the skills used in driving a car, require processes that "are unconscious but still work fast enough to control our real-time behaviours" Vold (2015, 25). Bodily factors certainly matter in occurrent cognitive states just as they matter in experiential states—if the body really acted as a low-pass filter, these cognitive processes would turn out to be internal in the same way as experiential processes. At least, for the reason that Clark is raising such a strong distinction between neural and non-neural processing, he should provide convincing reasons why cognition is different from consciousness—otherwise he is creating an objection against himself. Of course, this is not a knockdown to the high-bandwidth argument as such but is surely a reason why Clark should not want to use it. However, as I showed above, there are plenty of reasons why the argument based on high bandwidth is not a convincing way to debunk ECM in the first place.

5.3 Predictive Processing

Clark's second attempt at criticising ECM was published in his 2012 paper, where he answered Ward's (2012) answer to his 2009 paper, and came up with a new line of criticism towards ECM. This time, he appeals to *predictive processing* as a positive story how phenomenal consciousness is realised in the brain (alone). Clark claims that predictive processing is able to explain not only veridical, wakeful experiences but also explanatorily problematic experiences in dreams, hallucinations, and imagination. Clark himself, among other things, is famous for developing the theory of predictive processing: the branch he promotes is referred to as *action-oriented predictive processing* or *radical predictive processing*, and in general it is compatible with (at least some of) the 4E and situated views (see, e.g., Clark 2012, 2013, 2015, 2016).

The branch that Clark supports is contrasted with a moderate or conservative account of predictive processing that is committed to the view that generative power comes from full-blown (internal) representations (for overviews, see Clark 2015; Dolega 2017). However, as my interest here is not in the theories of predictive processing as such but rather in the counter-arguments against ECM, I will focus on the branch promoted by Clark, that is, action-oriented predictive processing. Once again, Clark has pioneered an entire research field to a new direction: the research in the intersection between 4E and predictive processing has become a very popular topic in philosophy of mind and cognitive science recently (see, e.g. Kirchhoff and Kiverstein 2019, 2020; Deery 2021; Gallagher and Allen 2016).

Predictive processing is a theory used in cognitive sciences, computational neuroscience, and other related sciences of the mind, concerning the generative power of the brain (see, e.g. Friston 2003, 2010; Wiese and Metzinger 2017). The brain, to put it bluntly, is a prediction machine whose task is to correct and minimise errors. The brain is generating and constantly updating a model of the environment based on the sensory information it receives. Prediction-driven processing is meant to reduce prediction errors, namely the difference between actual and predicted sensory signals. The world does not "come inside" the brain, so the brain predicts how the world is likely to be (Clark 2012, 759).

> To successfully represent the world in perception, if these models are correct, depends crucially upon cancelling out sensory prediction error. Perception thus involves "explaining away" the driving (incoming) sensory signal by matching it with a cascade of predictions pitched at a variety of spatial and temporal scales. These predictions reflect what the system already knows about the world (including the body) and the uncertainties associated with its own processing. [...] What we perceive depends heavily upon the set of priors (including any relevant hyper-priors) that the brain brings to bear in its best attempt to predict the current sensory signal. (Clark 2013, 187)

Clark's account is compatible with the externalist views. He himself states: "Embodiment and action fit very naturally within such a framework" (Clark 2012, 760). Embodied agents can actively control their sensory input stream, and that enables better and faster predictions. For example, turning one's head and walking towards an object are ways of controlling the sensory input, and hence the prediction process. Clark calls this "self-structuring of information flows" (a term borrowed from Pfeifer et al. 2007). The self-produced action creates a "known transform": I can move my head such-and-such ways to transform the incoming sensory data (Clark 2012, 760).

As we can see, Clark aims to come up with a view that would explain the positive results of sensorimotor enactivism and ECM (the features of sensorimotor enactivism that were presented in the second chapter, such as virtual presence/presence in absence and variable neural correlates)—while preserving an internalist account of conscious experiences. However, the most important asset for Clark (2012) is that he claims that the predictive processing framework can explain how perception-like states, such as dreams and hallucinations, resemble "normal" veridical perceiving phenomenologically, yet they don't have a direct link to the outside world. Clark claims that his account can explain the similarities and differences—in a way that supports the internalist explanation of the realisers of conscious experiences. Clark's idea is that in "normal" veridical perception, the generative models of predictive processing enable us to engage (at least to some extent directly) with the outside world. These same generative models are at play in the absence of the sensory information from the outside world, that is, in dreams and hallucinations. Accordingly, for Clark, perception and perception-like states have the same (internal) neural basis.

Hence, (this time) Clark rejects ECM because of the "super-tight link" between the mechanisms of perception and the mechanisms of phenomenally close states, such as dreams, hallucinations, and imaginings, and we should "consider these latter cases as co-arising within the general setting of mechanisms for veridical perception and effective action" (Clark 2012, 767). According to Clark, the predictive processing framework shows how we can have a genuine "not-indirect" access to the outside world—yet seal the mechanisms for perceptual experience in the head because the mechanism is the same as in (internal) dreams.

5.3.1 *Answering the Counter-argument from Predictive Processing*

My aim is not to criticise the action-oriented predictive processing framework itself. I think it provides a convincing story about how the brain anticipates its states and how this enables perception to reach a world that appears coherent. However, I disagree with Clark in that it would refute ECM. I claim that the predictive processing framework—as Clark presents it—is fully compatible with external material realisers of perceptual experience.

Michael Kirchhoff and Julian Kiverstein (2019) take a step further and not only argue that the predictive processing framework does not refute ECM but also argue that it provides further support for ECM. Hence, this is the altogether opposite of Clark. Instead of using the theory to criticise ECM, they use it to reinforce it—even though their reasons are based on the same empirical evidence. For them, prediction error minimisation is active, embodied engagement in the environment, and importantly, it is not unconscious. Organisms act to avoid unexpected or surprising sensory states through "active inference" that is real, online embodied action. According to Kirchhoff and Kiverstein (2019, 52), active inference has the consequence that "what we experience in perception is partly realised by what we do—by how we act—just as sensorimotor enactivism contends". That is, they read active inference in terms of sensorimotor enactivism. For them, predictive processing provides a way to explain the realisation part of phenomenal consciousness, whereas sensorimotor enactivism explains the character side.

It indeed seems that predictive processing does not pose an obstacle to ECM but quite the contrary offers a neuronal level explanation for the phenomenon described in ECM. Besides Kirchhoff and Kiverstein, others have also stressed the compatibility.[2] Ken Pepper (2014, 101) is on the same track and states that the predictive processing framework is "conducive to ECM and the enactive approach generally, as it puts the anticipatory structure of perception on a firm neurocomputational footing hitherto absent from the discussion". However, my aim here is more moderate: I do not aim to argue that predictive processing entails ECM, but I merely intend to show that it is *not incompatible* with ECM, and therefore we can disregard the counter-argument based on it.

Let us shift focus back to Clark's counter-argument. First, it is important to notice that Clark doesn't commit himself to an indirect view of perception—which would at least seem to be a quite straightforward way to deny ECM and set up internalism. If the world out there was just a model in our brains, the denial of the impact of environmental interaction would be much easier to establish. However, Clark's intention is that we don't perceive the generative models in our brains, but we perceive the world. "Equipped with brains like ours we become porous to the world.

[2] See, e.g. Deery (2021), who has argued that the proponents of the predictive processing framework have made an implicit endorsement that perception and imagination belong to the same mental *kind* that is a product of the same internal mechanism. He shows how this "common kind assumption" is not supported by the theory of predictive processing itself and argues that it is implausible. This leads to the conclusion that there is no in-principle barrier to hold ECM while holding the predictive processing framework.

Its structure and statistical regularities flow through us in as real a way as do food and water through the digestive tract" (2012, 767).

The directness of perception is related to the reason why Clark cannot consistently maintain his view. Pepper (2014) emphasises that if perception really is direct, then the phenomenological differences between veridical perception and dreams and other non-veridical processing are explained by the external elements in the world. He refers to O'Regan's notion of the world as an *external working memory*: "the stability of the waking perception is explained not in terms of sameness of representational content, but in terms of the stability and continual availability of information in the world itself" (Pepper 2014, 105). The external parts that are available in veridical perception, but absent in dreams, make a difference. I will discuss the dream question in more detail in Sect. 5.5.

The main reason why the predictive processing framework doesn't work as an objection to ECM is embedded in Clark's own philosophy. If we accept the *action*-oriented side of the theory that Clark himself promotes, ECM fits very well into the picture. As mentioned earlier, Clark (2012, 760) talks about "the self-structuring of information flows", namely bodily actions that are taken in order to obtain better and faster predictions. Clark's theory is embodied and enacted through and through, and therefore already permeated by consciousness. Pepper (2014, 102) has summed up the consequences in the following way:

> It is an unnoticed consequence of Clark's position that the EM theorist who endorses the [action-oriented predictive processing] framework cannot deny extended perceptual cognition on pain of inconsistency. This is because the centrality of informational self-structuring to [action-oriented predictive processing], when considered in light of some key commitments of EM expressed by Clark elsewhere, unwittingly commits the EM theorist to an extended physical basis of perceptual cognition.

Accordingly, if one accepts Clark's positions about EM and about action-oriented predictive processing, it follows that one cannot deny ECM. Throughout his other writings, Clark (see especially 2008) has emphasised that informational self-structuring is one of the exemplar cases of (occurrent) EM. How we position our bodies, turn our heads, make use of eye saccades, perform gestures, and so on should be seen as part of the cognitive processing itself—and not something that just affects, enables, or shapes it. There is a great deal of empirical evidence for this, for example Goldin-Meadow et al. (2001) and Goldin-Meadow (2003) studies on the role of gestures as part of the thinking process (for details, see Sects. 2.4.1. and 3.3.1.1.). Clark is famous for arguing for this kind of off-loading of the cognitive burden into the environment.

Pepper (2014, 102) points out that according to Clark's own characterisation of informational self-structuring (as sketched above), bodily actions have a constitutive role for the mechanisms of perception. Therefore, if one wants to hold on to EM (as Clark defines it) and also embrace the action-oriented predictive processing framework (as Clark defines it), one cannot just rule out the external elements—because otherwise EM would be cut out too. Therefore, Clark's criticism, as he presents it in his 2012 paper, does not work.

However, one might try a less direct form of criticism. As we saw, the direct rejecting of the role of the external elements in perception did not work out. But perhaps Clark would declare that he is only talking about the cognitive aspects of the mechanisms of perception but not the experiential aspects. For this claim, he would need an additional argument because the theory of the action-oriented predictive processing did not provide one. The high-bandwidth argument (i.e. consciousness requires faster processing power than cognition) would be an option, but as I showed earlier, it does not work. As I have argued elsewhere in this book (see especially Chap. 2, argument 1), the cognitive and conscious aspects cannot be separated as Clark has claimed. Clark's (2012) other alternative is the existence of dreams and hallucinations, and the "super-tight link" between them and veridical perceptions. I will discuss in more detail in Sect. 5.5 why this route doesn't provide the result that Clark wants.

To recap, the (action-oriented) predictive processing framework does not provide independent reasons to reject ECM. It needs further reasons for the now very familiar problem: how is one to uphold EM while denying ECM? As has now become clear, the task is very difficult, if not impossible, and in the best case the surviving EM would be a shrunken version that would cover only dispositional states and leave out the most noteworthy aspects that characterise Clark's theory of EM.

5.4 The Brain-in-a-Vat Thought Experiment

What if we removed some poor chap's brain from his head, placed it in a vat with nutrients, and simulated the inputs usually coming from the rest of the body and environment by wires to a supercomputer? Would this envatted brain have the same experiences (or at least some experiences) as a person with a normal brain situation? If that would be the case, it would prove quite efficiently that the brain is a necessary and sufficient material realiser for experiences.

The brain-in-a-vat is an archetypal example of a *thought experiment* in philosophy. Its origin can be traced back as far as Cartesian scepticism, and it has certainly provided food for thought for generations of philosophers.[3] However, a thought experiment as a method is problematic when applied to empirically oriented philosophy of mind. The reason for this is that whether a brain-in-a-vat scenario is conceivable, that is, metaphysically or conceptually possible *in principle*, doesn't yet have any impact on EM or ECM, or questions generally related to cognitive and neurosciences. Hurley (2010, 109) calls these kinds of arguments *supervenience thought experiments*, which "assume that internal and external factors do not vary together in relevant possible worlds—that internal factors can be unplugged from one array of external factors and plugged into another". However, if the internal factors turn out to be unpluggable, then the supervenience thought experiment is not possible in the first place (see Thompson and Cosmelli 2011, 173).

However, Thompson and Cosmelli (2011) (and see also Cosmelli and Thompson 2010) have approached the matter from an interesting point of view. It is interesting for two reasons. First, it very effectively nips in the bud the brain-in-a-vat counter-argument, and second, this refutation (i.e. the practical impossibility of envatting the brain) demonstrates that the body and environmental stimulation are necessary for the mind and consciousness. Therefore, they end up using the thought experiment for exactly the opposite purpose, namely for supporting the embodied, extended mind.

Thompson and Cosmelli make it clear that in order to have any argumentative power, the question must be addressed from a technical, bioengineering perspective. They follow Dennett's (1991) idea presented in a prelude to his book *Consciousness Explained*, where he asks what it would take to envat a brain so that it would function like a normal, embodied brain. Dennett's treatment shows that sometimes "*impossibility in fact* is

[3] For some early versions of the argument, see Dennett (1978), Putnam (1981), Smith (1984). Locked-in syndrome could be considered a "real-life case" of the brain-in-a-vat thought experiment. There are some enactivist answers about how to deal with the challenge the locked-in syndrome poses to 4E-views (see, e.g. Kyselo and Di Paolo 2013).

theoretically more interesting than a *possibility in principle*" (Dennett 1991, 4; quoted in Thompson and Cosmelli 2011, 168). First, the brain needs to be kept alive and protected by something that replaces the skull.

> [T]his system must involve at least the capacity to keep up with the energetic, ionic, osmotic, and recycling needs of the brain. It will therefore include some kind of circulatory system, plus the necessary pumps, oxygenating devices, and additional subsystems for ensuring the maintenance of physiological levels in the circulating fluid. [...] [W]hat the brain requires at any given instant depends on its own ongoing, moment-to-moment activity. Therefore, the life-sustaining system must not only be supportive of this activity, but also locally and systemically receptive and responsive to it at any given instant, independent of any external evaluation of the brain's needs. Consequently, to keep the brain alive and functioning, this responsive system will most likely need to be energetically open, and self-maintaining in a highly selective manner. (Cosmelli and Thompson 2010, 368–369)

Thus, this system needs to be robust and self-regulating—as organisms are. As Thompson and Cosmelli (2011, 170) note, this system begins to resemble more a body than a vat. And this is not enough. The thought-experimenter must also fix the following issues: brain's self-generated activity must be tightly coupled with the "body", and the stimulation that the nervous system usually receives from the body and environment must somehow be replaced. These are highly problematic requirements. First, it is not enough to try to mimic the normal inputs since identical inputs don't lead to identical states.

> It's not the case that an embodied brain and its envatted duplicate will remain qualitatively identical simply because they receive identical inputs throughout their lives. The brain isn't a reflex machine whose activity is externally controllable through input instructions. Rather, it's a highly nonlinear and self-organizing dynamical system whose activity exhibits an extreme sensitive dependence on initial conditions. Inputs perturb such complex systems, but don't specify particular outcomes. Furthermore, most inputs arise as a consequence of the system's own intrinsic activity. Hence to get the body-type inputs to match the normal inputs precisely would require getting them to match the bodily inputs to the brain that arise from the brain's nonlinear and unpredictable intrinsic activity. (Thompson and Cosmelli 2011, 171)

The thought-experimenting engineer should find out how to produce phenomenal experiences that would match those of the normal experiencers. As we learnt from the previous passage, this is not a simple task. Rather, it is very probably an *impossible* task. Some kind of stimulating device would be needed that would be "in perfect synchrony with the brain's exploratory motor efference signals as its sensory systems scan through the virtual image, and updating its activity so as to match precisely the sensory reafference" (Thompson and Cosmelli 2011, 172). Further, the stimulating device should not disturb the life-sustaining system: every stimulation creates a disturbance in the homeostasis that needs to be compensated by the system (Thompson and Cosmelli 2011, 172; Dennett 1991, 5).

Studies in computational neuroscience[4] indicate that adaptive behaviour rises from the dynamical coupling and feedback between the nervous system and the peripheral sensorimotor systems—it is not a one-directional process programmed by the brain. Hence, the "vat" should have the required flexibility and capability to ensure this dynamical coupling relation. As Thompson and Cosmelli (2011, 172) point out, "the best and probably only way to establish this coupling is to give our brain real sensorimotor systems it can control". Hence, the brain in its vat looks very much like an embodied agent in the world.

Therefore, *envatting* the brain actually requires *embodying* the brain. It doesn't have to be identical with our body, but it needs to fulfil certain functional requirements,[5] that is, it has to be a *surrogate body*. This makes the thought experiment impossible, because—if we recall Hurley's notion of *supervenience thought experiment*—the factors (brain vs. body) were meant to be separable for the purposes of the thought experiment. Nevertheless, they are not. The brain is not a brain (as we know it) without its body and environment. The brain is explanatorily and physically unpluggable from the body.

As I mentioned at the beginning of this chapter, Clark (2009a, 981) explicitly rejects the brain-in-a-vat argument: if the vat doesn't imitate everything the real world provides, the thought experiment is not fair, and

[4] Thompson and Cosmelli refer, for example, to a study made by Chiel and Beer (1997), where it is shown that feedback loops between the nervous system, body, and world are essential for behaviour. See also, e.g. Ferri et al. (2012) and Chiel et al. (2009).

[5] This leaves the door open for the possibility of an *artificial mind* (e.g. a robot). Thompson and Cosmelli, however, do not address the questions of developmental robotics and artificial intelligence.

if it does, it is unable to prove anything, since it is impossible to separate what is the role of the substituting external interactions. Clark's reasoning is close to what I have been arguing so far, and what I think is the main reason why the brain-in-a-vat is not a good argument: if the "vat" doesn't mimic everything, the experiment isn't fair—and if it does, the result is a surrogate body situated in a surrogate environment.

Moreover, the idea of a disembodied brain creating experiences by itself is in general very far from the enactivist and 4E-framework. Dave Ward illustrates this with two comparisons. A man performing movements identical to tangoing on his own doesn't qualify as dancing a tango, since the relation with a partner and music are necessary for something to count as dancing a tango. Similarly, "driving" an overturned car is just spinning its wheels and not driving, since the act of driving requires a relationship between the car and the road (or another surface). The same goes for brain activity and experiencing. "A disembodied brain going through a sequence of physical states in limbo is not experiencing, for it is in no meaningful relationship with its environment" (Ward 2012, 746–747). Disembodied legs are not walking, and a disembodied brain is not experiencing because the necessary parts of the picture are missing.

Finally, even if the brain-in-a-vat scenario were possible, it would only speak for that specific brain in that specific vat, and it might still be that with real embodied people in natural situations the experiences were constructed in the interactions with their bodies and environment. To recap, this counter-argument is unsuccessful all the way down the line and is of no threat to the externalist views. Somewhat surprisingly, despite all this, I think the brain-in-a-vat argument still manages to reveal something. Its impossibility illustrates that "the brain and body are so deeply entangled, structurally and dynamically, that they are explanatorily inseparable" (Thompson and Cosmelli 2011, 168). In order to maintain a mind, the brain simply cannot be unplugged from the body. Body states, brain states, and environmental interactions are inseparable, and belong to the constitutive basis of experiences.

5.5 Dreams and Hallucinations

The counter-argument based on the existence of dreams and hallucinations is perhaps the most straightforward (and most obvious, in a sense) that has been presented against ECM. The central idea is similar to the brain-in-a-vat argument, but with one striking difference. Whereas the brain-in-a-vat scenario leaned on a thought experiment, this argument is based on a common everyday phenomenon, of which there is an abundance of empirical research available. I think that of all the counterarguments presented in this chapter (also including the causal–constitution fallacy presented in Chap. 3) the dream argument is the most convincing—even though I do think the ECM supporter can answer it. However, it is striking how little literature there is about this topic, namely the relationship between 4E-theories and dreams. Antti Revonsuo (2015) has contributed a direct objection to ECM and sensorimotor enactivism based on the "dream argument". Enactivists have not given a proper reply. This section aims to fill that gap—or at least raise the central issues we need to take into consideration in order to respond to the challenge.

The dream argument is the following. If conscious experiences are (partly) realised by external elements as ECM claims, how are states such as dreams and hallucinations possible, as it seems that they are realised exclusively in the head? If dreams and hallucinations do not need any interaction with the environment, possibly other experiences do not either. And if the interaction with the environment is not a necessary feature for these certain experiences, how can it be part of the constitutive basis for experiences in general? We need to accept the situation described in the following passage that appears to be challenging the very core principle of the externalist view of consciousness. "The dreamer is highly conscious (she has vivid experiences), is disconnected from the environment (she is asleep), but somehow her brain is creating a story, filling it with actors and scenarios, and generating hallucinatory images" (Nir and Tononi 2010). This is the challenge that dreams and dream-like states posit for ECM and externalist views in general.[6]

I will focus on dreams, since there is more literature about them and because dreaming is a very familiar phenomenon, of which (almost) all of us have first-hand experience. Hallucinations, on the other hand, are considered to be something that is not so common (this might be a false belief, but still, it is fair to say that dreaming is a more mundane and commonplace phenomenon than hallucinating). However, the existence of

[6] At least externalist views that argue for the extension of experiential states. ECM is not the only one of this kind, for example, sensorimotor enactivism (and perhaps enactivism in general) and the hypothesis of extended affectivity (see, e.g. Colombetti 2014) are in the target. However, I think dispositional EM is in the clear when it comes to the dream challenge (though some might oppose this based on the fact that we do have beliefs in dreams).

hallucinations (whether due to schizophrenia, brain injury, psychedelic drugs, alcohol, sleep deprivation, etc.) would be an equally good example. I sometimes use the term dream-like states, by which I refer to dreams, hallucinations, illusions, imaginings, and other non-veridical experiencing.

There is one crucial methodological matter we need to take into account. Philosophers who are addressing the dream question very often tend to ignore all the empirical research there is about dreaming. Matthew Ratcliffe (2017) has made a sharp observation. He points out that philosophers are using their very own conception of hallucinations (he calls it *philosophers' hallucinations*) that has been in a dominant position through the history of philosophical thought experiments, as Descartes's evil demon and various more recent counterparts show. This kind of concept of hallucination, however, has barely any connection to another concept of hallucination, that is, *real hallucinations*, that is studied in psychiatry. This is a very important remark: we need to be aware in which sense the concept is used because the meaning is very different. Ratcliffe's sharpness notwithstanding, he has a similar kind of blind spot himself when it comes to *dreams*. There is a similar discrepancy between *philosophers' dreams* that are used in thought experiments and *real dreams* that empirical dream research is investigating. Philosophers often use a notion of dreams that has no connection to empirical research.

Revonsuo (2015) has strongly criticised the fact that the results of empirical studies are often absent from philosophers' theories. I agree with him that when philosophers of mind and cognitive science define dreams, it is crucial that they take into account the empirical results from neurosciences and dream research. However, I disagree with him that these empirical results would necessarily entail an internalist account of the mind or of phenomenal experience.

Another problem is that the writers who have addressed the dream challenge regarding ECM have a strong dissonance even about the very starting point, namely what the relation between dream experiences and veridical experiences is. The opinions vary from one end of the spectrum to the other. According to Revonsuo (2015), dreams are phenomenologically similar or even identical with veridical experiences—they simulate waking life. According to Pepper (2014), dreams are somewhat different phenomenologically—they are an impoverished version of veridical experiences and waking life. According to Noë (2009), dreams are completely different—they belong to a different *kind* than veridical experiences. This reveals that the question is complicated because this is only the starting point that does not even address externalism or internalism yet.

It is good to remember that of these three writers, only Revonsuo is a dream researcher. Considering this, it is surprising how strongly he appeals to the similarity or identity between veridical and dream experiences, when after all, besides the similarities, there is also plenty of empirical evidence of *dream bizarreness*, such as discontinuities, holding contradictory beliefs, sudden transformations, and impossible scenes or activities, like pink elephants or flying (Nir and Tononi 2010; Hobson et al. 1998). Moreover, as I will argue in Sect. 5.5.2.4., Revonsuo's interpretation or understanding of *phenomenal experience* is different from the notion that sensorimotor enactivism and ECM deal with.

Even though this dissonance has complicated the discussion, I claim that the nature of dream experiences and their level of similarity with veridical experiences are not actually of the essence concerning the question we want to address here. For the dream argument to work, it is enough that *some* dreams are similar *enough*. Kristjan Loorits (2017, 123) calls this "sufficient similarity": a dream is sufficiently similar to veridical experience when the dream experience contains the phenomenological features that are explanatorily relevant for veridical experiences. And there is a consensus amongst the dream researchers that there are dreams that belong to this group (see, e.g. Revonsuo and Salmivalli 1995; Domhoff 2007).

Further, even though for the sake of the (opponent's) argument I will treat dream experiences as fully internally constituted processes, this in fact is not necessarily true. The material realisation of dream experiences is perhaps not restricted to neural processes only, for it might also depend on bodily functions. As Metzinger (2013) states, the *functional disembodiment hypothesis*, according to which dream states are disconnected from the sleeping body (see Windt 2015), has been refuted by empirical studies. Instead, there is evidence that some minimal forms of embodiment are preserved during REM-sleep dreams. Inputs from the real sleeping body can enter dreams, and vice versa, dream movements may be enacted by the sleeping body. According to Metzinger, the most evident case is the "reliable directional correspondence between dream-eye movements and real-eye movements". Moreover, studies in lucid dreaming indicate that dream movements are in accordance with muscle twitches in the same limbs as in the dream and also correspond with heart and respiration rates. External, real-body stimuli, such as spraying water on the skin and electric stimulation, have also been shown to be mirrored in our dream bodies (Metzinger 2013).

These results show quite unquestionably that dreams are not completely disembodied. However, the embodiment question would require a more thorough investigation, which I am not going to pursue here. I settle for the modest claim that it is not necessary that dream experiences are entirely disembodied. Moreover, even though the evidence speaks for some amount of embodiment, it doesn't indicate *extension* outside the body. The potential minimal embodied nature notwithstanding, I will treat dream experiences as purely brain-bound events. A proponent of ECM can answer the dream challenge either way.

In what follows, my aim is to show that empirical dream research doesn't contain anything that would necessarily lead to an internalist account of the mind. First, I will introduce the most straightforward dream argument against ECM that has been presented in the literature and discuss some of its merits and deficiencies. I will then discuss four different options how a proponent of ECM can defend herself from the dream challenge in general.

5.5.1 Revonsuo's Dream Argument

Revonsuo (2006, 2015) has set up the most elaborate version of the dream argument against ECM and sensorimotor enactivism.[7] According to him, if the following two statements are true, internalism of consciousness follows, and he claims that "all the available evidence from empirical dream research" supports them. *First*, dream experiences are phenomenally highly similar and, in most cases, identical to the veridical waking-life experiences—they have equally rich and complex experiential contents, and this is manifested especially during REM sleep. *Second*, if the first clause is true, and dream experiences are as rich and complex as the waking experiences, it follows that the only minimally necessary constitutive conditions for experiential states are the ones that are active in the brain during dreaming in REM sleep[8] (Revonsuo 2015, 57–58).

I am hesitant to accept even the first premise. It is undeniably true that waking and dreaming experiences are phenomenally similar to a great

[7] There are some differences how a proponent of sensorimotor enactivism and a proponent of ECM can answer the challenge, though. I will elaborate these differences in the following sections.

[8] As a technical matter, we might ask how empirical dream research could offer even in principle any support or resistance for the second statement, which aims to be a deductive inference, and not something that could be verified or falsified by empirical means.

extent, but empirical dream research doesn't back up the claim that they would have equally rich and complex experiential contents (Hobson et al. 1998). Further, I disagree with the second premise and the conclusion (the brain-boundedness of dream experiences leads to the general brain-boundedness of *all* experiences, and therefore an internalist theory of consciousness follows). The jump from dream experiences to all experiences cannot be taken in as straightforward a manner as Revonsuo assumes. I will discuss the reasons for this below, and my suggestions for answers to the dream argument (presented in following sections) deal with this very same question.

Even though I disagree with Revonsuo's conclusion and find his argumentation wanting in various ways, I do agree that he addresses an important matter—a challenge that is pressing for ECM and deserves an answer. It is valuable as such to open the discussion and by so doing demand an answer from the enactivists. The core of the challenge is perhaps best captured in Revonsuo's (2015, 58) elaboration. Compare a sleeping person in dreamless sleep at T1 with the same person at T2 in REM sleep and vivid dreaming. In the first case the explanandum is completely absent, whereas in the second case it is present—while everything in the dreamer's environmental relations (or the lack of them) stays constant. Therefore, Revonsuo continues, as the only material differences are the neural differences between T1 and T2, they have to be constitutively responsible for the explanandum.

This shows why a *prima facie* obvious answer that is sometimes heard from 4E-minded people (see, e.g. Noë 2009, 180) is insufficient. It would seem natural to suggest that dreaming might as well happen completely in the head, but it wouldn't be possible without all the previous environmental interactions we have had during our lives and over hundreds of thousands of years of evolution. Even though this "obvious" answer is not a straightforward solution, it is in line with the "explanatory strategies", such as *weak enactivism*, which I will present later as one possible solution to the dream argument. However, this answer does not address the issue Revonsuo is posing with the dream argument. Internalists (such as Revonsuo) would probably accept the aetiological role of external relations in the (causal) explanation of dreams. As I will show soon, there are also ways to answer the argument that are based on constitutive strategy.[9]

[9] This said, my aim is not to undermine the causal, aetiological relation that past waking-life experiences and actions have on dreams or their evolutive role.

Revonsuo's aim is to criticise sensorimotor enactivism in general, but he only refers to one writer's one book (i.e. Noë 2009). Moreover, the section where Noë discusses dreams is a short one. I mainly agree with Revonsuo's critique regarding Noë's view (I will discuss Noë's view in Sect. 5.5.2.1.). However, a few pages from one book (which also happens to be written for a wider audience) from one author is not a comprehensive sample of sensorimotor enactivism. Loorits (2017) has elaborated Revonsuo's arguments and stressed that sensorimotor enactivism contains different positions and facets regarding dreams, such as Hurley's (2010) *explanatory* strategy, according to which dreams require a brain-bound explanation, even though with veridical experiences, a sensorimotor explanation is more apt. Another example is a skill-based view, supported, for example, by Myin (2016), that stresses the counterfactual element—skills, capabilities, and dispositions to act, rather than online action—in the constitution of perceptual experiences.[10] Hence, Revonsuo's refutation of Noë's short passage is not enough to prove the whole of sensorimotor enactivism wrong.

Loorits (2017) agrees with the conclusion Revonsuo draws: the existence of internally constituted dream experiences offers us reasons to believe that veridical experiences are also constituted internally. However, he also criticises Revonsuo's straightforward argumentation in the two statements presented above. Revonsuo claims in his second premise that the brain-bound nature of dream experiences leads to the brain-bound nature of all experiences, including veridical waking-life perceptual experiences. This is too hasty a conclusion, and dream research does not offer any support for it. As Loorits (2017, 111) notes, the empirical evidence Revonsuo refers to only shows that dream experiences and veridical experiences are phenomenologically similar. The conclusion Revonsuo draws is not the only possible consequence from this empirical fact. For example, one could equally well say that dream experiences and waking experiences are realised in a different manner.

Even though the straightforward critique Revonsuo is using is not possible, it is true that dreaming shows that sensorimotor enactivism cannot pass the *general theory of consciousness* (Loorits 2017, 107), that is, a theory that comprises all areas of consciousness. Revonsuo's argumentation rests on passing the general theory. According to him, if a theory "cannot

[10] I will call this view *weak enactivism* and discuss whether it is of any help with the dream challenge in Sect. 5.5.2.3. See also the discussion about *weak* vs. *strong enactivism* in Chap. 2.

explain dreaming in particular, then it cannot explain consciousness in general" (2015, 53). He seems to be under the impression that any kind of experience without the environmental relation will refute the externalist theory. However, this is not a necessity, and there is no principled reason why a theory of consciousness should be applicable to all cases of consciousness. To sum up, I can grant Revonsuo that dream experiences are brain-bound states that are phenomenologically sufficiently similar to veridical experiences. However, I disagree that this would lead to an internalist account of veridical (perceptual) experiences. In the next sections, I discuss different options to argue for this view.

5.5.2 *Answering the Dream Challenge*

A proponent of ECM can try to answer the dream challenge in at least four ways. *First*, it is possible to deny the connection between dream experiences and veridical experiences altogether. If dream experiences are not real experiences (or don't belong to the same *kind* with veridical experiences), obviously their existence doesn't pose any threat to externalist theories. *Second*, we can look at the definition of experience itself. If it is understood in an enactivist way rather than as static *what-it-is-like-ness*, the question changes. *Third*, we can appeal to a narrower interpretation of sensorimotor enactivism that stresses mastery of skills, and knowledge or understanding of sensorimotor contingencies rather than actual interactions between the subject and her environment. The skills may stay intact during dream experiences, and the core of enactivism prevail. *Fourth*, we can accept that experiences don't have to extend in all cases. In the spirit of *multiple realisability*, it is possible to hold a view that certain veridical waking-life perceptual experiences have an extended base, whereas some other instances of experiences, such as in dreams, hallucinations, and imaginations, are brain based.

In the following sections, I elaborate on the benefits and deficits of these different options. I will argue that the first option is problematic, mainly because it contradicts current dream research. Further, I admit that the second option might not make an opponent of externalism satisfied because it presumes a conception of experience that is already extended. Perhaps this option should be read more as a descriptive statement than as an argument against the dream challenge. However, I think it still deserves its place because it reveals the significance of the conceptual background, as well as fundamental differences between the externalists and internalists. However, the third and fourth options are the best ways to surmount the challenge.

Noë's Disjunctive Enactivism

One option is to deny that veridical and dream experiences share the same phenomenological status. If dreams are not real experiences (because their phenomenality is different or lacking), the way they are constituted (internally or externally) is irrelevant for ECM. Alva Noë (2009)[11] supports this kind of *disjunctivist*[12] view in order to promote his sensorimotor approach. He admits that unlike veridical experiences, the seemingly similar states we have during dreaming are not extended. This is not, however, a problem, since he holds that dream experiences are not real experiences to start with. Dreams and dream-like states only *seem to be* similar with veridical perceptual experiences, but in fact they belong to different *kinds*. Going through what appears as "an experience of red" during a dream is phenomenologically different from a veridical waking-life experience of red, and only the latter is a real instance of an experience of red. Dreaming is less rich in details, "dream seeing" is not "real seeing", and all dream experiences are based on earlier veridical experiences.

I call Noë's view disjunctivist enactivism. His view is disjunctivist because what we call 'experiences' are either (1) real, veridical perceptual experiences or (2) quasi-perceptual experiences that belong to a different kind. His view is enactivist because real experiences are based on an agent's online actions in the real environment.[13] Thus, Noë can hold a view according to which all (real) experiences are extended without needing to pursue the somewhat wild claim that experiences would be extended during dreams.

Noë's view is partly problematic. The main problem is the core claim that dream experiences are not real experiences. This claim not only is unintuitive but also contradicts dream research—especially experiments with lucid dreamers indicate that dreams really are conscious experiences (see, e.g. LaBerge et al. 1981; Windt 2015; Revonsuo 2015; Domhoff

[11] Noë has also discussed the dream question briefly elsewhere (see, e.g. 2004, 2008).

[12] Disjunctivism rejects the view that experiences have a common nature: veridical, dream, hallucination, and so on; experiences belong to different kinds. They might appear similar, but the similarity is only ostensible. For disjunctivism in epistemology of perception, see, for example, Martin (2002).

[13] Disjunctivist enactivism is my term; neither Noë nor Revonsuo call it a disjunctivist view.

2007). Even though dream experiences are not identical in every sense, they clearly fulfil the definition of experiential states. They are *sufficiently similar*, as was stated in earlier sections. Noë also claims that dreams merely consist of experiences we have experienced previously (in waking life). This claim is not supported by empirical studies (Revonsuo 2015), and even if was true, it wouldn't back up the externalist account since the causal background conditions are not what the dream argument is referring to anyway. Hence, the concessions we would need to take with Noë's account are too serious to accept, and we can conclude that disjunctive enactivism cannot answer the dream challenge in a satisfying way.[14]

Experience as Sense-Making

Even though I found Noë's account wanting, I think this underlying statement is valid: dreams are not experiences *if* we understand and define experiences as enactivists do. Let us observe what kind of concept of experience the dream challenge is dealing with. It seems that when Revonsuo speaks about experiences, he refers to fixed states of *what-it-is-like-ness* (reminiscent of qualia). However, if we understand experiencing as process-like *sense-making* that is partly constituted by the very interactions the experiencer has with her environment, the setting changes. The enactivist notion of experience contains an extension to start with: the organism–environment system that brings about conscious experience is decomposable. When experience is defined in this way, dreaming is obviously ruled out. Dream experiences don't fulfil the definition of experience that enactivism is endorsing, and hence the problem is solved.[15]

This kind of enactivist line drawing is of course a rather controversial way to answer the dream challenge because it depends on our previous definition of the explanandum. Thus, it is a question-begging claim—when the question is whether dream experiences are experiences. Thus, an opponent of ECM probably would not be convinced with this option to answer the challenge. Of course, we can take this option as a descriptive rather than an argumentative point.

[14] Apart from Noë's account, there have been earlier attempts to claim that dreams are not experiences (Malcolm 1959; Dennett 1976). Their argument was that all we know about dream experiences derives from the moment of waking up, so perhaps we never had the experiences but instead only false later memories of them. However, this view has been proved wrong by empirical dream research (see, e.g. Revonsuo 2006).

[15] Perhaps this was what Noë had in mind in his 2009 book, but he doesn't explicitly state this, so I don't claim that this is Noë's view.

After all, as I have shown in other parts of this book, we have other, independent reasons to support the enactivist account of (some) experiences. As I have argued, experience is (partly) constituted by the skilful active relation with the environment. Especially sensorimotor enactivism that was presented in the second chapter stresses these features. Qualitative differences are explained by *sensorimotor contingencies*—the understanding of how our movements and actions affect our surroundings and objects in lawful ways. Our sensorimotor understanding of the world and the sense-making relationship created in that interaction constitute the perceptual experience—it simply *is* the experience. When that relationship is absent—as in dreaming—experiences are absent. Admittedly, when dreaming, we certainly undergo processes that resemble sensorimotor experiences, but because the necessary constitutive element is missing, dreams cannot belong to the phenomenon that sensorimotor enactivism is describing. They are experiential states, but not those dynamic interactions we have with our environment. Noë (2009, 179–180) writes:

> [D]ream seeing is not really seeing at all. [...] [W]e ought to think of perceiving as an activity of exploring the environment. It is not a process whereby a picture of the world is built up in your brain; rather, it is the activity whereby you achieve access to what is around you by making use of various different skills (of movement, of understanding, etc.). [...] The content of our experience—what we experience—is the world; in the world's absence, we are deprived of content. For this reason, whatever we think and feel and say, when we dream, we are not seeing.

Revonsuo (2015, 61) criticises Noë for question-begging in his argumentation. It is true that Noë *presupposes* an externalist account of experiences and does not truly face the challenge the dream argument poses. His opponent is no better when it comes to presuppositions. Revonsuo (2015, 56) writes: "I am here assuming an *internalist* theory of consciousness: consciousness is located inside the brain, and thus if there is any constitutive mechanism for consciousness, it must also reside in the brain". How could a dialogue be possible between two completely opposite views that are based on strong presuppositions? It seems that neither of these writers can step out of their own background assumptions, where dream experiences are concerned. The following two options for answers (viz. weak enactivism and multiple realisability of experiences) are not based on presuppositions about conscious experience. In the next section, I will argue that enactivists can also answer the dream challenge without begging the question.

Weak Enactivism

As it has now become clear, the dream argument shows that the online active explorative relationship with the environment is not necessary for there to be (some kind of) conscious experience. Environmental interactions are simply absent during dreams—we don't probe, explore, touch, reach, or move with our sleeping bodies[16]—and hence a sensorimotor theory that depends on these relations cannot be applied to dreams. What does this mean for sensorimotor enactivists? Do we have to either commit to sensorimotor enactivism and deny that dreams belong to its scope, or embrace internalism and deny sensorimotor enactivism? Fortunately, there is a third option. Sensorimotor enactivism can also be understood in a way that doesn't require that the online relation is active in order to explain experiences. I am talking about *weak enactivism*, which was already discussed in the second chapter (see Sects. 2.3.3. and 2.3.6. and Aizawa 2010a).[17]

According to weak enactivism, our *knowledge* of how we can access the world and how our actions affect it (namely knowledge of sensorimotor contingencies) is all that is needed for constituting experiences. As opposed to *strong enactivism*, the weak version abandons the requirement for an actual online bodily movement—the disposition to act is enough. In order to perceive a coffee cup as three dimensional, we don't have to walk around it because we possess a knowledge (which of course originates from earlier online interactions) what it would very probably look like from other sides and how its appearance would change were we to observe it from different positions. For weak enactivism, possessing this knowledge is sufficient.

Of course, these skills and understanding of sensorimotor contingencies are already embedded in the actual interactions—the features described by weak enactivism are embedded in strong enactivism, but the feature

[16] Of course sleepwalking, talking, and various other activities (*somnanbulism*) sometimes occur, but they are connected with low conscious, non-rapid eye movement (NREM) sleep stages. However, usually during dreaming, the body is in a sort of paralysed state called REM muscle atonia (Szelenberger et al. 2005).

[17] Loorits (2017) describes a somewhat similar account (which he then attempts to refute). He refers to Myin (2016) and Beaton (2013) as advocates of this kind of view.

specific to strong enactivism (i.e. online action) is not necessary. Strong enactivism explains not only (waking-life) actions, such as reaching for a coffee cup, but also perceptual experience without action, such as looking at a coffee cup. Nevertheless, even in the latter case (looking at a coffee cup in a wakeful state), there is undoubtedly an online relation and access to the world—the cup is really there, there is real visual information coming from it to the retina, and perceivers know how their potential actions would affect the cup. With veridical perceptual experiences, I think strong enactivism is the best explanation, and there is no need to apply the weaker version, not even with apparently more "passive" experiences, such as having a perceptual experience of a blue coffee cup (i.e. colour perception).

However, the situation might be different with dream experiences. Looking at a coffee cup in a dream is different. It cannot be explained by our *actions*, as was the case in reaching for the coffee cup in waking life. Neither can it be explained by an online access to the coffee cup, because there is no coffee cup. Hence, strong enactivism clearly cannot explain dream experiences. In the absence of actions and online access to the world, what is left? Even during dreaming, some of our capacities and skills remain intact. The understanding of sensorimotor contingencies that we have gained through our lives in the actual interactions with our environment doesn't disappear altogether. We can reach for a dream coffee cup (and manage to get it in our dream hand or not) or walk around a dream house. We use the same skills in the dream environment as we use in waking life. The actions might have different and unexpected results, but our mastery of certain basic sensorimotor contingencies remains to a certain extent similar. Moreover, we don't even have to move the dream cup (in the dream situation) because the possibility or disposition to move it is enough to count as mastering a skill. Thus, I suggest that weak enactivism could offer a way to *explain* dream experiences.

This should be taken as an explanatory strategy rather than as an ontological or constitutive statement. The skilful understanding of sensorimotor contingencies plays an explanatory role in dream experiences. On the one hand, it explains the *similarities* with veridical experiences by appealing to the same skills we use in online perception. On the other hand, it explains the *differences* from veridical experiences by appealing to the lack of actions and online access to the world. I think this is the closest we can get to an enactivist explanation of dreams. In the next section, I will discuss an even more straightforward explanation that is not dependent on enactivism.

Multiple Realisability of Experiences

All the options so far were dependent on committing ourselves to stronger or weaker versions of enactivism. We have yet another option that is independent of the previous views and independent of enactivism. This answer is to simply admit that experiences are not extended during dreams, but that this is not a problem for ECM, since for ECM to be a sound theory, it is not necessary that all experiences are always extended. As opposed to Noë's view, this option is not to claim that dream experiences belong to a different kind than veridical experiences. Rather, it appeals to the possibility of the *multiple realisability* of experiences. Hence, internally constructed dreams don't necessarily entail internally constructed veridical experiences because veridical perceptual experiences can be constituted in one way and dream experiences in another. They can be very similar, maybe they can even be phenomenally indistinguishable from each other but still have a different material basis.

For example, Hurley appoints dream-like states in the brain-bound sphere and veridical perceptual experiences in the sphere of environmental interaction loops. (She calls the phenomenon *variable neural correlates* rather that multiple realisability, but the idea is the same.) According to her "neural correlates can vary despite sameness of quality", and this happens across veridical experiences as well as between veridical and non-veridical experiences (Hurley 2010, 119).

For this option, it is irrelevant whether the dream experiences and veridical experiences are phenomenally identical, somewhat similar, or entirely different from each other. The crux is that their material realisation can differ either way—even if they are phenomenologically identical—if we accept the possibility of multiple realisability. This is obviously at least a logical possibility, but would it be possible to solve empirically? Can we show empirically that somewhat similar experiences can have a different material realisation? There is plenty of evidence that veridical experiences can be constituted in different ways. One example is *neural plasticity*. It provides an empirical example of how different brain areas can produce similar phenomenological states. The use of inverted goggles is also an example where different material realisations produce the same phenomenal experience before and after adaptation to the distortive lenses (Hurley 2010, 119–120). Psychiatric disorders could be another example. For example, according to some philosophers of psychiatry, depression can be constituted in various different ways (see, e.g. Borsboom et al. 2019).

Unfortunately, dream research doesn't provide unambiguous results because the studies have found both similarities and differences at phenomenological and neurophysiological levels of explanation when comparing dreams and veridical experiences (for an overview, see Nir and Tononi 2010). It is uncontroversial that waking-life cognition enables and delimits dreams. On the other hand, some phenomenological differences (the ways in which dreaming experiences differ from waking-life experiences) are accompanied by neurophysiological differences. Subjects' cognitive and neural organisations are in accordance with an important degree in waking life and dreaming. For example, children's dreams develop consistently with their (waking-life) cognitive capacities: small children's dreams are mostly simple and static. Patients with brain lesions have similar deficiencies in their dream experiences as in their waking-life experiences, for example, patients with impaired face recognition do not dream of faces (Nir and Tononi 2010; Domhoff 2003; Foulkes 1999).

All in all, empirical questions regarding dreams are far from settled. Nir and Tononi (2010) note that not even all dreams are constructed in the same way neurophysiologically, and thus researchers should treat "different kinds of dreams and their neural correlates separately". However, it seems that empirical neuroscience cannot use multiple realisability as a general methodological tool—otherwise it would be impossible to make generalisations across species (see, e.g. Bechtel and Mundale 1999). The multiple realisability argument is far from unproblematic, and my aim is not to claim that we need to commit to a strong metaphysical version of it in order to save ECM from the dream challenge.

Rather than committing to strong metaphysical claims, my intention with multiple realisability and the empirical cases supporting it has simply been to show that experiences don't have to be realised in the same way, but instead their material constitutive basis can vary significantly. Dream research provides evidence for this. "In REM, the brain [...] becomes activated but processes its internally generated data in a manner quite different from that of waking" (Hobson et al. 1998, 1). During waking, brain activity is modulated by aminergic systems, while in REM sleep it is modulated by cholinergic systems (Hobson et al. 1998; Pace-Schott and Hobson 2002). Thus, it seems that the constitutive material basis differs (or at least it is not identical) between waking experiences and dream experiences.

A visual experience of a red apple can be constituted (partly) externally, and a dream experience of a dreamed red apple can be constituted entirely internally—and this is not a problem for my position. Most of the proponents of externalist views tend to agree that an all-encompassing extension is not necessary. Chemero and Silberstein (2008, 130) have phrased it clearly:

> Extended cognitive scientists need not claim that all of cognition is extended, just that some of it is. There is no *a priori* reason why some phenomenon might be given wholly brain internal explanations (e.g., dreams, certain kinds of pain), some given embodied accounts, while others will require explanations that span brain-body-environment systems (e.g., motor control, long division). Indeed, we all ought to embrace causal and explanatory pluralism about cognition.

Conclusions from the Dream Argument

I claim that the reason why the dream challenge cannot refute ECM is a combination of the last three options I described above. The most important reason is simply the following: for ECM to be a sound theory, it is not necessary that all kinds of experiences are extended. A supporter of ECM doesn't need to deny that dreams are real experiences, for empirical dream research has shown quite undeniably that they are genuine experiences. However, this is not a problem for ECM, as I have shown.

It is clear that a strong version of sensorimotor enactivism cannot explain dream experiences. A proponent of that view can justify this in two ways. On the one hand, it is possible to say that dream "experiences" are not experiences at all or state that the enactivist concept of experience doesn't include dream experiences by definition. On the other hand, if we are ready to accept the weaker version of enactivism, dream experiences can also be explained within the theory.

The dream argument would certainly work against a theory that claims that all experiences are always extended. However, I don't think many people would commit themselves to this even amongst the strongest of enactivists. Even if they did, the same would not apply to ECM because ECM is *not* a theory about all experiences, but about a limited group of perceptual experiences. Ward has put into words the principal answer to the dream challenge and criticisms of similar kind (even though he is not talking about dreams). "What is required to rule out ECM is a reason to believe that neural states alone are *always sufficient* for the subpersonal basis of experience" (Ward 2012, 743).

Of course, yet one more straightforward (or simplifying) answer would be to say that because ECM is a theory about *perceptual* experience, and dreams are not perceptual experiences, their existence has nothing to do with ECM. However, this answer would bypass the challenge altogether, because its gist is to bring up that there are experiences that are very similar to the experiences that externalist theories are trying to explain but which are constituted entirely internally. I think the proponents of ECM and other externalist views should rather welcome this kind of testing and trying out of their theory—for it offers a fine opportunity to improve and elaborate some questions that might otherwise have gone unnoticed.

5.6 Summarising the Counter-arguments

This chapter presented the most important arguments against extended experience and offered suggestions how the challenges can be responded to. First, I showed that Clark's two arguments against ECM (based on high bandwidth and predictive processing) are unable to refute ECM. Interestingly, they share the same problem: were they true, they would not only refute ECM but also Clark's own theory of EM. This observation supports one of my main claims in this book: it is very difficult to argue for EM and deny ECM at the same time.

Both of the last two counter-arguments (brain-in-a-vat and dreams and hallucinations) rest on the idea that if we can show that there are some cases of internally constituted experiences, it follows that ECM is wrong. I showed especially with the dream argument that this argumentation is not sound. The same reasons can be applied to the brain-in-a-vat objection, too. However, I stressed different aspects in my answers. With the brain-in-a-vat argument, I focused on the general implausibility of the whole scenario, whereas with the dream argument I did not challenge the internal nature of dream experiences but rather the internalist conclusion that the opponent of ECM can draw from it.

References

Adams, F. & Aizawa, K. (2001): "The Bounds of Cognition". *Philosophical Psychology*, 14(1): 43–64.

Aizawa, K. (2010a): "Consciousness: don't give up on the brain". *Royal Institute of Philosophy Supplement*, 67, 263–284.

Beaton, M. (2013): "Phenomenology and embodied action". *Constructivist Foundations*, 8(3), 298–313.

Bechtel, W., & Mundale, J. (1999): "Multiple Realizability Revisited: Linking Cognitive and Neural States," *Philosophy of Science*, 66, 175–207.

Borsboom, D., Cramer, A., & Kalis, A. (2019): "Brain disorders? Not really: Why network structures block reductionism in psychopathology research". *Behavioral and Brain Sciences*, 42. https://doi.org/10.1017/S0140525X17002266

Chalmers, D. (2008): "Foreword". In Clark, A.: *Supersizing the Mind: Embodiment, Action and Cognitive Extension.* Oxford, UK: Oxford University Press.

Chalmers, D. (2017): "Extended Cognition and Extended Consciousness". In Colombo, M., Irvine, E., & Stapleton, M. (eds.), *Andy Clark and his Critics.* Wiley-Blackwell (forthcoming).

Chemero A. & Silberstein M. (2008): "Defending Extended Cognition". B. Love et al: Proceedings of the 30th Annual Meeting of the Cognitive Science Society. Austin, TX.

Chiel, H. & Beer, R. (1997): "The brain has a body: Adaptive behavior emerges from interactions of nervous system, body and environment". *Trends in Neurosciences*, 20, 553–57.

Chiel, H., Ting, L., Ekeberg, Ö. & Hartmann, M. (2009): "The Brain in Its Body: Motor Control and Sensing in a Biomechanical Context". *Journal of Neuroscience*, 14, 29(41), 12807–12814.

Clark, A. (2008): *Supersizing the Mind: Embodiment, Action and Cognitive Extension.* Oxford, UK: Oxford University Press.

Clark, A. (2009a): "Spreading the joy? Why the machinery of consciousness is (probably) still in the head". *Mind*, 118(472), 963–993.

Clark, A. (2012): "Dreaming the whole cat: Generative models, predictive processing, and the enactivist conception of perceptual experience". *Mind*, 121(483), 753–771.

Clark, A. (2013): "Whatever next? Predictive brains, situated agents, and the future of cognitive science." *Behavioral and Brain Sciences*, 36 (3), 181–204.

Clark, A. (2015): "Radical predictive processing". *The Southern Journal of Philosophy*, 53, 3–27.

Clark, A. (2016): *Surfing uncertainty: Prediction, Action, and the Embodied Mind.* Oxford: Oxford University Press.

Colombetti, G. (2014): *The Feeling Body: Affective Science Meets the Enactive Mind.* Cambridge, Massachusetts: The MIT Press.

Cosmelli, D. & Thompson, E. (2010): "Embodiment or Envatment? Reflections on the bodily basis of consciousness". In J. Stewart, O. Gapenne & E. Di Paolo (eds.), *Enaction: Towards a New Paradigm for Cognitive Science.* MIT Press.

Deery, J. (2021): "Extending the extended consciousness debate: perception, imagination, and the common kind assumption". *Phenomenology and the Cognitive Science*. https://doi.org/10.1007/s11097-021-09738-x

Dennett, D. C. (1991): *Consciousness Explained*. Boston: Little Brown.

Dennett, D.C. (1976): "Are dreams experiences?" *Philosophical Review*, 73, 51–171.

Dennett, D. C. (1978): "Where Am I?" In *Brainstorms: Philosophical Essays on Mind and Psychology*. Cambridge, MA: MIT Press, 310–322.

Dolega, K. (2017): "Moderate Predictive Processing". https://doi.org/10.15502/9783958573116

Domhoff G.W. (2003): "The Scientific Study of Dreams: Neural Networks, Cognitive Development, and Content Analysis". *American Psychological Association*.

Domhoff G.W. (2007): "Realistic simulation and bizarreness in dream content: Past findings and suggestions for future research", in Barrett, D. & McNamara, P. (eds.) *The New Science of Dreaming*, vol. 2, 1–27, Westport, CT: Praeger.

Eliasmith, C. (2008): "Dynamics, Control, and Cognition". In Aydede, M., & Robbins, P. (eds.), *The Cambridge Handbook of Situated Cognition*, Cambridge University Press, 134–54.

Ferri F., Frassinetti F., Ardizzi M., Costantini M. & Gallese V. (2012): "A sensorimotor network for the bodily self". *Journal of Cognitive Neuroscience*, 24: 1584–95. https://doi.org/10.1162/jocn_a_00230

Foulkes D. (1999): *Children's Dreaming and the Development of Consciousness*. Cambridge, MA: Harvard University Press.

Friston, K. (2003): "Learning and inference in the brain". *Neural Networks*, 16 (9), 1325–1352.

Friston, K. (2010): "The free-energy principle: A unified brain theory?" *Nature Reviews. Neuroscience*, 11 (2), 127–138.

Gallagher, S., and Allen, M. (2016): "Active inference, enactivism and the hermeneutics of social cognition". *Synthese*, 195: 2627–2648.

Goldin-Meadow, S. (2003): *Hearing Gesture: How Our Hands Help Us Think*. Cambridge, MA: Harvard University Press.

Goldin-Meadow, S., Nusbaum, H., Kelly, S. D., & Wagner, S. (2001): "Explaining math: Gesturing lightens the load". *Psychological Science*, 12(6): 516–522.

Hobson J. A., Stickgold R, & Pace-Schott E. (1998): "The neuropsychology of REM sleep dreaming". *Neuroreport* 9: R1–R14. https://doi.org/10.1097/00001756-199802160-00033

Hurley, S. (2010): "The Varieties of Externalism". In Menary, R. (ed.), *The Extended Mind*. Cambridge, MA: MIT Press, 101–153.

Kirchhoff, M. D. & Kiverstein, J. (2019): *Extended Consciousness and Predictive Processing: A Third Wave View*. London, UK: Routledge.

Kirchhoff, M. D. & Kiverstein, J. (2020): "Attuning to the world: The diachronic constitution of the extended conscious mind". *Frontiers in Psychology*. 11:1966.

Kyselo, M., & Di Paolo, E. (2013): "Locked-in syndrome: a challenge for embodied cognitive science". *Phenomenology and the Cognitive Sciences* 14(3): 517–542.

La Berge, S. P., Nagel, L. E, Dement, W. C., & Zarcone, V. P., (1981): "Lucid dreaming verified by volitional communication during REM sleep". *Perceptual and Motor Skills*, 52(3), 727–732.

Loorits, K. (2017): "Dreaming about Perceiving: A Challenge for Sensorimotor Enactivism". *Journal of Consciousness Studies*, 24 (7–8), 106–129.

Malcolm, N. (1959): *Dreaming*. Routledge & Kegan Paul, London.

Martin, M.G.F. (2002): "The Transparency of Experience". *Mind and Language*, 17, 376–425.

Metzinger, T. (2013): "Why are dreams interesting for philosophers? The example of minimal phenomenal selfhood, plus an agenda for future research". *Frontiers in Psychology*. https://doi.org/10.3389/fpsyg.2013.00746

Myin, E. (2016): "Perception as Something We Do". *Journal of Consciousness Studies*, Volume 23, 5–6, 80–104.

Nir, Y. & Tononi, G. (2010): "Dreaming and the brain: From phenomenology to neurophysiology". *Trends in Cognitive Sciences* 14(2), 88–100.

Noë, A. (2004): *Action in Perception*. Cambridge, MA: MIT Press.

Noë, A. (2008): "Magic Realism and the Limits of Intelligibility: What Makes Us Conscious?" In Hawthorne, J. (ed.) *The Philosophy of Mind*. Oxford: Blackwell. 457–476.

Noë, A. (2009): *Out of Our Heads*. New York: Hill & Wang.

Pace-Schott, E. & Hobson, A., J. (2002): "Basic mechanisms of sleep: New evidence on the neuroanatomy and neuromodulation of the NREM-REM cycle". In Davis, K., Charney, D., Coyle, J., & Nemeroff, C. (eds.) *Neuropsychopharmacology: The Fifth Generation of Progress*. American College of Neuropsychopharmacology.

Pepper, K. (2014): "Do sensorimotor dynamics extend the conscious mind?" *Adaptive Behavior*, 22(2), 99–108.

Pfeifer, R, Lungarella, M., Sporns, O., and Kuniyoshi, Y. (2007): "On the information theoretic implications of embodiment- principles and methods". *Lecture Notes in Computer Science*. 4850, 76–86.

Putnam, H. (1981): "Brains in a Vat". In *Reason, Truth, and History*, Cambridge: Cambridge University Press, 1–21.

Ratcliffe, M. (2017): *Real Hallucinations: Psychiatric Illness, Intentionality, and the Interpersonal World*. London and Cambridge, MA: The MIT Press.

Revonsuo, A. (2006): *Inner Presence: Consciousness as a Biological Phenomenon*. Cambridge, MA: MIT Press.

Revonsuo, A. (2015): "Hard to See the Problem?" *Journal of Consciousness Studies*, 22 (3–4), 52–67.

Revonsuo, A. & Salmivalli, C. (1995): "A content analysis of bizarre elements in dreams" *Dreaming*, 5 (3), 169–187.
Smith, P. (1984): "Could We Be Brains in a Vat?" *Canadian Journal of Philosophy* 14, 115–123.
Szelenberger, W., Niemcewicz, S. & Dabrowska, A. (2005): "Sleepwalking and night terrors: Psychopathological and psychophysiological correlates". *International Review of Psychiatry*, 17(4), 263–270.
Thompson, E. & Cosmelli, D. (2011): "Brain in a Vat of Body in a World? Brainbound versus Enactive Views of Experience". *Philosophical Topics* 39(1), 163–180.
Vold, K. (2015): "The parity argument for extended consciousness". *Journal of Consciousness Studies* 22: 16–33.
Ward, D. (2012): "Enjoying the spread: Conscious externalism reconsidered". *Mind*, 121(483), 731–751.
Wiese, W. & Metzinger, T. (2017): "Vanilla PP for Philosophers: A Primer on Predictive Processing". https://doi.org/10.15502/9783958573024
Windt, J. M. (2015): *Dreaming: A Conceptual Framework for Philosophy of Mind and Empirical Research*. Cambridge, MA: MIT Press.

CHAPTER 6

Concluding Remarks and the Future of Extension

In this book I have shown in many ways that one cannot consistently hold an externalist view about the material basis of cognition while denying it regarding consciousness. If one endorses extended cognition, one has to endorse extended consciousness as well. However, many philosophers have tended to overlook this. I presented several reasons why phenomenal experience cannot be omitted. In the beginning, I introduced the current externalist framework in philosophy of mind and cognitive sciences and placed my own position at the top of third-wave externalism. I also outlined the central concepts, accounts, and differentiations that were used and developed in the later chapters.

In the second chapter, I argued that if we accept the general 4E-framework, we have to accept ECM as well. The arguments included bridges from the hypothesis of extended cognition (i.e. Clark's & Chalmers's EM) to ECM, from sensorimotor enactivism to ECM, and finally from the embodiment thesis to ECM. There is a way to understand the first (EM) and third (embodiment) premise theories without endorsing ECM. However, if this is done, it requires that the premise theories be significantly restricted. I argued that sensorimotor enactivism differs in this sense: it cannot be accepted without endorsing ECM as well.

In the third chapter, I addressed a subject matter that has gained a lot of attention in recent research on both the philosophy of mind and the philosophy of science, namely the *causal–constitution* debate. I labelled

P. Telakivi, *Extending the Extended Mind*, New Directions in Philosophy and Cognitive Science,
https://doi.org/10.1007/978-3-031-35624-7_6

"the constitution-turn" the immense interest in the ontology of constitution as a determining question of whether EM and other externalist views are true. The abundance of literature started with Adams and Aizawa's early critique of Clark and Chalmers's EM, and later, it spread to and was mingled with accounts in the philosophy of science. Whilst using the *new mechanist approach* as a starting point, I argued that rather than in mechanist terms, the causal–constitution separation should be interpreted in *diachronic–dynamical* terms when we are dealing with mental phenomena. When understood in this way, the extension relation in ECM counts as *constitutive*. The discussion in this chapter was also a way of responding to two central critiques against ECM, namely the objections based on *the causal–constitution fallacy* and *the cognitive bloat argument*.

In the fourth chapter, I set the preconditions and limits for my position of ECM. I started by distinguishing between four degrees of extension: *momentary extension*, *integrated extension*, *prosthetic incorporation*, and *functional incorporation*. The last one is required for ECM: the external tool has to be part of the transparent bodily point of view and under the sensorimotor control of the subject. Based on the notion of functional incorporation, I formulated the *glue and trust* criteria for ECM. They include *quasi-transparency*, *sensorimotor control*, *ownership*, *sense of agency*, and *sense-making attitude*. Finally, I discussed a phenomenon that fulfils my glue and trust conditions, for which there is a great deal of empirical research available, namely *sensory substitution* technology.

Finally, in the fifth chapter, I discussed four counter-arguments that have been set forth against ECM: (1) *the argument from high bandwidth*, (2) *the argument from predictive processing*, (3) *the brain-in-a-vat thought experiment*, and (4) *the dream challenge*. The last two concern the same theme: if we can have phenomenal experiences without online interactions with the environment, does that show that internalism also prevails with perceptual experiences? With the former objection, I concentrated on showing how the brain-in-a-vat thought experiment is not on firm ground to start with, because it is based on a highly dubious science-fiction scenario. With the latter objection, I presented different ways to answer the dream challenge and concluded that a strong version of (sensorimotor) enactivism cannot explain dream experiences. My main solution to the dream argument was simply that it is not necessary that all kinds of experiences are always extended for ECM to be a valid theory about certain perceptual experiences. Internally constructed dreams do not lead to internalism in the case of veridical experiences. Veridical perceptual experiences can be constituted in one way and dream experiences in another.

This book is exceptional within the 4E-framework because it combines extended functionalism with an enactivist view. This kind of unifying contribution has been lacking from both functionalist accounts of the extended mind and enactivist accounts of the mind. It was necessary to begin this enterprise from a functionalist starting point because that is the theoretical framework of Clark and Chalmers's EM. However, I hope the reader has followed my lead throughout the book towards a more enactivist understanding of cognition and consciousness. Once we have taken this path, there is no need to return to the functionalist point of departure.

Another asset of this book is that it combines empirical sciences with the philosophy of mind, which is not always done in philosophy. As I wrote in the first chapter, there has been an appeal for this kind of broader account, since a comprehensive analysis of ECM has been lacking from the literature. Alongside Kirchhoff and Kiverstein's (2019) book, this is the first study to directly address ECM in book length.

However, the work is not over. This book raised several questions for future research, for example regarding the relationship of the predictive processing framework and ECM, the status of dream experiences, and the role of developmental robotics in testing the sensorimotor accounts of the mind. Furthermore, this research might provide tools for related areas that were not directly addressed in it. Even though this book focused on extended perceptual experience, I think some of the results can be applied in questions regarding the potential extension of *affectivity*, *self*, and *social interactions*. After all, they all share the experiential side, although the topics otherwise differ.

The question concerning the borders of the mind bears direct implications for many social, ethical, and legal questions. How should we interpret the legal status of a device that functions as part of the mental system? Perhaps hacking a laptop or stealing a smartphone should be seen as an assault on one's person rather than on one's property. It is also possible that extending one's mind is not always empowering, but it could lead to cognitive atrophy. As new technologies become more and more immersed in our minds and societies, these questions will be amongst the most important for the humankind.

Another important ethical outcome concerns people whose resilience has suffered. The role of external resources becomes more obvious with people in vulnerable situations, such as elderly people with memory disorders or people who suffer from psychiatric disorders. If we accept the extended view, we should pay more attention to external objects, spaces, places, and social relations when deciding on diagnoses and treatment. In this sense, it seems that an externalist view of the mind could be more ethically viable than an internalist view, since it gives a stronger status to certain environmental elements as part of someone's set of mental capacities.

Yet another interesting topic for future research is the use of AI and machine learning technology as an extender (see Hernández-Orallo and Vold 2019). What if instead of a "traditional" tool such as a notebook, the external part is a device based on AI technology? For example, AI-based tools (such as apps and wearables) can start manipulating their users in ways that are not possible for basic tools like notebooks. On the other hand, they can also offer very advanced and personalised assistance; for example, as aids for patients with dementia, autism spectrum disorders, or learning disorders—and even create completely new capacities.

In his newest book about virtual and augmented reality, David Chalmers (2022) returns briefly to the olden days. Instead of 53rd Street, New York, he has now landed in Sydney. He introduces to us two people, named Omar and Ishi, who both want to visit the Opera House. Ishi is not technologically enhanced, so she ponders for a moment, recalls the route, and walks there. Omar, on the other hand, is a dedicated user of augmented reality glasses, and he relies on them for everything he does. He utters the word "Opera House" and his glasses show him the way, and he walks under those instructions to the destination. Chalmers doesn't analyse the similarities and differences between basic and AI extenders, but undoubtedly there looms an enormous field of new research questions ahead of us. Fortunately, we have the theories of extended mind and extended conscious mind to guide us.

Clark and Chalmers's opening question was "*where* does the mind stop and the rest of the world begin?" However, as we have seen throughout this book, the *where*-question is not as straightforward one might first think. It should not be interpreted as a question about the *fixed location* of the mind. In contrast, the focus on location is downplayed within the 4E-framework. Instead, the boundaries, and hence the location, are negotiable and open to change. What counts as boundaries of the mind varies depending on what kind of environment we are in and what kind of opportunities and affordances that environment offers. What matters is what we do and how we exploit the resources. Whether a blind person's cane or a forgetful person's diary is part of their minds depends on how the devices are used and whether they are integrated or incorporated into their users.

References

Chalmers, D. (2022): *Reality+: Virtual Worlds and the Problems of Philosophy*. UK: Allen Lane.

Hernández-Orallo, J. & Vold, K. (2019): "AI Extenders: The Ethical and Societal Implications of Humans Cognitively Extended by AI". *AIES'19*, January 27–28. Honolulu, HI, USA.

Kirchhoff, M. D. & Kiverstein, J. (2019): *Extended Consciousness and Predictive Processing: A Third Wave View*. London, UK: Routledge.

References

Metzinger, T. (2013): "Why are dreams interesting for philosophers? The example of minimal phenomenal selfhood, plus an agenda for future research". *Frontiers in Psychology*. https://doi.org/10.3389/fpsyg.2013.00746

P. Telakivi, *Extending the Extended Mind*, New Directions in Philosophy and Cognitive Science,
https://doi.org/10.1007/978-3-031-35624-7

Index[1]

[1] Note: Page numbers followed by 'n' refer to notes.

P. Telakivi, *Extending the Extended Mind*, New Directions in Philosophy and Cognitive Science,
https://doi.org/10.1007/978-3-031-35624-7